HOW IT FEELS TO BE
ATTACKED BY A SHARK

HOW IT FEELS TO BE ATTACKED BY A SHARK

And Other Amazing Life-or-Death Situations!

Michelle Hamer

SKYHORSE PUBLISHING

Copyright © 2007 by Michelle Hamer

All Rights Reserved. No part of this book may be reproduced in any manner without the express written consent of the publisher, except in the case of brief excerpts in critical reviews or articles. All inquiries should be addressed to Skyhorse Publishing, 555 Eighth Avenue, Suite 903, New York, NY 10018.

www.skyhorsepublishing.com

10 9 8 7 6 5 4 3 2 1

Library of Congress Cataloging-in-Publication Data

Hamer, Michelle.
How it feels to be attacked by a shark : and other amazing life-or-death
situations / Michelle Hamer.
p. cm.
ISBN-13: 978-1-60239-191-8 (flexibound : alk. paper)
ISBN-10: 1-60239-191-2 (flexibound : alk. paper)
1. Life change events. 2. Adjustment (Psychology) 3. Stress (Psychology)
4. Psychic trauma. 5. Crisis intervention (Psychiatry) I. Title.

BF637.L53H36 2007

155.9'3—dc22

2007026151

To Rob

I still can't believe my luck

CONTENTS

Introduction ix

How it feels . . .

to be attacked by a great white shark   1
to be struck by lightning—twice   6
to rescue a child from a flooded drain   10
to lose a leg   17
to weigh five hundred pounds   22
to choke to death on a cheeseburger   28
to be caught in an avalanche   33
to save a life on your wedding day   37
to die and come back   42
to win the lottery   46
to be crushed in an ice crevasse   51
to be mauled by a rottweiler   55
to be suicidal   58
to be shot in the head   63
to foil a midair hijack attempt   68

to survive a plane crash	72
to survive bacterial meningococcal disease	77
to be lost at sea	82
to be run over by a tractor — twice	87
to drown	92
to contract dengue fever	97
to be attacked by a crocodile	101
to be abducted by aliens	106
to be an animal psychic	114
to be "locked-in" (to be in complete paralysis)	118
to be an identical triplet	122
to be kidnapped and tortured	128
to be drug-addicted and homeless	137
to lose everything in a fire	144
to be shot in the heart with a nail gun	150
to have a triple organ transplant	154
to have quintuplets	160
to be caught in a cyclone	166
to battle cancer	172
when there is a murder in your family	178
to be brainwashed by a cult	184
to set yourself on fire	193
Notes	199
Acknowledgments	201

INTRODUCTION

You're in the jaws of a huge animal, the life force is being squeezed from your body and its immense teeth are ripping open your flesh . . .

Your lungs are exploding; blood begins to seep from your eyes and you know that if you don't get air soon it'll be too late—then the world goes black . . .

There's an explosion, it feels as if you've been hit in the head with a baseball bat and then you realize that blood is pouring from a head wound—you've been shot . . .

REMEMBER THE SHOCKING footage of lawyer Gerald Curry being shot five times at point-blank range outside a courthouse in California which brought violence and horror right into our living rooms? Or how about the strength and determination of star surfer Bethany Hamilton who survived a vicious tiger shark mauling that saw her lose her arm. And how many of us wondered if we too could have had the same gritty courage and determination as Aspen mountaineer, Aron Rolston who severed his own arm with a pocketknife to escape a 1000-pound boulder that was pinning him down.

All of these stories stay with us long after the headlines have faded because they are powerful reminders of what human beings can achieve when pushed to the limit; and that even in the face of unimaginable pain and suffering there is hope—and faith.

These sorts of intense experience happen to only a few of us, but capture all our imaginations.

If you've ever wanted to know just how it feels to experience some of life's more extreme moments then settle down and enjoy this collection of stories about ordinary folk who have achieved the extraordinary and who faced with crisis, were able to triumph.

These are the gripping, real-life stories of people who have teetered on the brink of death, faced life-changing experiences or found incredible inner resources will amaze and inspire you.

You'll be stunned by the sheer determination and endurance of each person in this book, and you'll wonder—what would I do in a similar situation? How would I cope? Could I be so brave? These stories will shake you to your core and remind you of the fragility

INTRODUCTION

of life—and of humankind's amazing ability to overcome even the worst that life dishes up.

You'll read about such incredible experiences as:

- The man who felt the jagged teeth of a shark crack against the bones in his legs.
- The policeman who risked his life to save a child from a flooded drain.
- The construction worker who accidentally shot himself in the heart with his nail gun and sat waiting to die.
- The woman who was death rolled by a crocodile—three times—and came out of the experience with a whole new take on life.

These are truly awe-inspiring stories, but you'll also feel humbled by how these everyday people, faced with extraordinary situations, triumph over potential tragedy and then move on to integrate the experience into the rest of their life.

Then there are lighter stories to enthrall readers, including the animal psychic who tunes into pets and tells their owners what food they like and what they really think about wearing that rhinestone collar. And the stories that are almost too bizarre to be true, such as the young man who forces a cheeseburger down his

INTRODUCTION

throat as a party trick, but then finds himself suffocating and perilously close to death. Or the shocked man who wakes to find himself being studied by aliens.

This book is packed with dozens of compelling first-person accounts along with stunning true tales to get your adrenalin pumping, your heart racing and to restore your faith in the power of the human spirit. Read these accounts and you will be drawn into a confronting reality that few of us, fortunately, will ever have to experience

It's been my privilege to walk a short way with each of these people on their journey through life and to collect their thoughts and memories. I'm sure you'll enjoy getting to know them too.

HOW IT FEELS TO BE
ATTACKED BY A SHARK

> *"When I felt his teeth hit my bone,
> I thought he was going to break me."*

HOW IT FEELS TO BE ATTACKED BY A GREAT WHITE SHARK

ALLAN OPPERT, 46

I'VE BEEN DIVING all my life. I love diving for crays, or rock lobsters. If the weather's good, we dive once a week. I've seen sharks in the water before, but it has never been a problem. I'd been in twice that day and gotten some crays, but I thought I'd go back for these last three. I jumped in the water feetfirst; normally I go in headfirst, but for some reason that time I didn't. I stopped at about fifty-five feet (that's nearly seventeen meters), and as I looked at my depth gauge, I was sort of in a standing position, just to let everything thicken up—my blood and so on.

As I studied my gauge, I looked down below me, and that's when I saw the great white. I thought, "Oh my

HOW IT FEELS . . .

God, here we go." Right away, I thought, "What a big shark!" Turns out he was about fifteen feet. I wasn't worried at that stage. I concentrated on slowing my heart right down. As soon as I identified what it was, I thought, "Well, a great white. If it comes at me, he's going to try to knock me out," because that's what they do—and he did exactly that.

I took a couple of deep breaths and relaxed and relaxed—because they can hear your heartbeats, these sharks—and I just thought, "Well, if it comes, it comes, but I'll be ready." And sure enough he spun around on the bottom and came straight at me. I took a deep breath and pushed my guts out.

As he was coming up he just opened up his mouth and punched me with his nose. Then both my legs were in his mouth and there was just a huge force, a huge push. It felt like my dive bottle was trying to come through my back because he was pushing me up to the surface with tremendous force. He hit me in the stomach with his nose, then he bit down, and I could feel the bite go in on top of my knees and underneath my calf muscles.

It was just excruciating pain. I thought, "Hang on, he's going to bite my legs in half, my knees are going to snap." The pressure was just unbelievable. I felt his teeth sink in and go right to the bone. I could feel his teeth grinding against my bones. My legs were in his mouth and he started thrashing me around like a rag doll—like a little fox terrier thrashing a rat.

It was the worst pain I've ever experienced—I couldn't wish that on anybody. That bite down on my knees and the bite at the back of my calf muscles—oh boy. You've got to remember that, in the bottom of its jaw, the great white has two sets of teeth. When I felt his teeth hit my bone, I thought he was going to break me. I was reaching for my knife to poke him in the eyes, but I think that as he came up, his teeth just sheared the knife off the side of my leg, and then when he bit down, I think he bit into my spear gun, which lay across my lap. I'd put a new head on that spear gun and this thing was really sharp.

My mask had come off the side of my face, so I pushed it back on and I'm looking down and thinking, "Whoa, what a big head." His eyeballs were, I think, about thirty-five inches apart. He was twisting me around so much that I could see the gills just down below his eyes and I could count them all. This probably lasted ten seconds—or ten hours, I couldn't say.

He had my legs in his mouth, and I was face-to-face with this massive head. I was thinking, "This is not happening, this is not happening, but yes, it's happening." I knew I was going to die, but I didn't think it would be at that particular moment. He let me go—he spat me out, I think. Once he'd bitten through the top of my knees, he'd also bitten through my spear gun and bitten it in half, and three-quarters of it had gone down his gullet. My theory is that my knife was inside him already—and it was a brand new knife too—and my spear gun's down there too, and he probably thought, "This is no good," and he let go.

By then my BCD was full of air. That's my buoyancy control device, which you wear as part of your vest, but I can't remember pushing the button to inflate it. It's compulsory when you're diving deep. I was being pushed up to the surface. He let me go and I went blasting straight out of the water; I probably went the last three or four meters just flat out. My mask had slipped down over my chin on the way up, so you can imagine the force.

When I got to the surface, that's when I thought I was going to die. As I got there, just knowing what these great whites do, I knew I had about ten seconds before he came back to finish me. I was counting, "A thousand-and-one, a thousand-and-two . . ." waiting for him to come after me.

I yelled out to the boys on the boat to come and get me, that I'd been attacked by a shark. I was screaming at them. As soon as they began to pull me into the boat the shark came up underneath me. One of my friends yelled out, "Hurry up, hurry up, he's here! He's coming up!" They pulled me in, and he came straight up underneath our boat and straight across to the other boat and attacked the propeller. He was pretty mad; he had a go at the anchor as well. He wasn't very happy at all.

My injuries were pretty horrific. I knew how bad they were because I'd probed with my fingers. I had to put my kneecap back together. It was sticking up in the air so I pushed my fingers in there and pushed it down. I could feel the muscles and tendons spilling out onto my legs. The pain was extreme; it was worse when I bound my legs up with an old fishing jersey and a towel. That's when my feet started to go numb.

It took us about forty minutes to get back in. I was lucky not to bleed to death. I could feel the blood running down and pooling around my buttocks. There was plenty of blood. I was in the hospital for six days, and then it was about three weeks before I could walk again. I don't have any feeling in the back of my calves now, and sometimes my knees seize up.

I've been back diving several times since, but not at that depth. I will, though, because that's the only way to get the big crays, and I love getting them. But I'm more wary now; I tend to look around a bit more carefully. I look into the distance as far as I can. I never used to have a care in the world under the water, but now I'm more aware.

> *"It arced out and hit the water and just blew us all up in the air."*

HOW IT FEELS TO BE STRUCK BY LIGHTNING— TWICE

Don Whitford, 57

I'VE BECOME A bit of a standing joke because I worked for the Weather Bureau and I've been hit by lightning twice. The first time was while I was working in a town called Darwin. I was at soccer training, I went up for a mark at the goal post, and lightning hit the soccer ball. There was just a loud bang and it was quite frightening. I fumbled the ball and fell to the ground and we called training off.

It wasn't really a day of thunderstorms, so it was very hard to tell where it came from; it just happened unexpectedly. There was this sudden flash that zapped across from somewhere and hit the ball, just as I tried to put my hands on it. Strangely enough I didn't feel anything much, other than a tingling sensation, because of the fact that I wasn't grounded; I was up in the air. Somebody said I had a yellow glow around my body as the lightning hit the ball, but I didn't see anything. It was quite extraordinary. The lightning probably hit the goal post and then arced across from the goal post to the ball.

Afterward I felt a bit shaky, but I had no visible injury or pain. I could have been seriously injured if I was grounded, because the charge would have gone right through my body. I know that I was really very lucky. There're plenty of documented cases of people being severely injured, losing fingers and toes, or being killed. But we just laughed and thought no more about it really—it was all over in a flash—literally.

The second time was also in Darwin, about three or four years after the first time I was hit—and I really should have known better. I was running with a group called the Hash House Harriers, just a group of guys who got together to run during the week. This day the cloud base was about fifty feet from the ground. It was a really low, horrible, tropical thunderstorm, with lightning and thunder and torrential rain. We were running across flat ground with water up around our ankles—that's how heavily it was raining.

There were about twenty of us or so, running across an open oval, up across a little rise, and we spread out to go around this little tree which had just been planted—

it was no more than six and a half feet high—and then lightning struck. We think it hit the tree, but it arced out and hit the water and just blew us all up in the air, all of us. That was rather frightening and painful.

Immediately after it hit I knew what was happening and I just thought, "You idiots, what are we doing out in this?" I just cursed our stupidity for being there and I honestly thought somebody was going to be dead. I looked around and there were about twenty grown men in fetal positions on the ground, just lying still. It was quite horrifying. I thought: "Oh God, somebody's going to be dead here," but fortunately nobody was.

Four of us were quite badly off; we suffered burns and a bit of shock, and I had burnt feet because the charge going through the water boiled it and scalded our feet. It was as if someone had poured a boiling kettle over our feet. The skin was red-raw for a while. Somebody else had a big burn on his chest, another guy had a burn on the side of his head, and one guy stopped breathing for a while. An ambulance came and took four of us to the hospital.

I've had a whiplash injury in my neck ever since then, which is from being thrown up in the air. I just remember being airborne for a second or two, and it felt as if somebody had hit me over the back of the neck with an iron bar. As the volts go through you, your whole body jars and you go into spasm and your neck jolts back. It wasn't till we left the hospital that I thought about being struck twice, and I decided: "Well, strike three and I'm out; I'll have to be careful."

I stayed in Darwin for another few years after that, but after several more near misses, I made sure I got well out of it whenever there were storms around. One particularly horrific time we were out prawning, and as we walked out into the water, dragging the prawn net behind us on the mud flats, a storm came over and lightning hit the mud flats. Of course, we were the tallest thing on the mud flats, so we lay down in the mud, which was just lovely! Then it rained so much it was almost like the tide coming in. Then the mud crabs started to come up, and we thought, "This is like a horror movie," because lightning was hitting the ground within 150 feet of us. Eventually we just ran and hid. I think I've pushed my luck a bit with lightning.

After all those incidents I became pretty wary of storms. One time in Darwin I was mowing the grass and a storm came over, but because of the noise of the mower, I didn't hear it. All of a sudden there was a flash and a bang and I literally dropped everything and ran under the house in fear—it was just an automatic reaction. A few minutes later I felt like a fool and looked around to see if anyone was watching. It must have looked funny, but I just don't take any chances these days.

> *"It was terrifying.
> I honestly thought I was going to drown."*

HOW IT FEELS TO RESCUE A CHILD FROM A FLOODED DRAIN

Alan Sykes, *46*

IN 1996 WE had three days of continuous torrential rain in a place called Coffs Harbour, which resulted in severe flooding around the area. I was a detective in the police force at the time. One morning my partner Gavin Dengate and I went back to the station after an early job. We were just grabbing a cup of tea when we heard the call come over the radio that there was a young boy lost, possibly trapped in a stormwater pipe.

We raced around to the people who had made the phone call. There was a guy there with two little guys, and they were pretty traumatized. They had been playing in the flooded creek on their boogie boards, then were heading home when they turned around and real-

TO RESCUE A CHILD FROM A FLOODED DRAIN

ized this young guy who had been with them had just gone. They realized he must have been swept into a storm-water drain, which was only about thirty-five inches wide and in full flood. He wouldn't have been able to see anything because it would have become pitch dark very quickly.

This poor boy of only about eleven or twelve was washed with phenomenal force down this pipe and—who knows if it's divine intervention or not—the end of the pipe dips down at quite a severe angle and then goes into a grate, but somehow a piece of timber had become wedged across the pipe, just before that dip. In the pitch black he hit this thing and grabbed it. He was there, in the freezing cold, with a phenomenal force of water pouring on top of him, and all he could do for the next hour or so was just scream for his life.

We didn't know where he was, so we stripped off and pulled some cement covers from the drains, which allowed us to see down into the storm-water pipe. We thought we could see the boy, or what we thought to be his corpse, as there was no sound, no movement. It was about three hundred feet down the pipe. If it was him, we hoped we might be able to get him out and resuscitate him.

We didn't really have any equipment, and we were working against time because there'd been so much rain but, in another case of divine intervention, it had stopped raining by the time we got there. Somebody brought a rope. There was a bit of a verbal tussle between my partner and I about who was to go in there, but I was the senior man and I said I was going. If I had waited a second longer, Gavin would have been straight in there. We're both dads with little kids.

They gave me a rope and I tied it around my waist and jumped in. The flood of water washed me down about sixty-five feet. I shouldn't have done that because I realized immediately that the rope was totally and completely inadequate and that it wouldn't hold me. The water was just so strong. I'm a pretty big guy, about six foot two and 240 pounds, but this water was just washing me around like a feather—it was incredible. I said, "Stop this, get me out," so they pulled me out.

By that stage the SES (State Emergency Services) had arrived. They gave us a proper lifeline rope. More cops had arrived. I said I'd go down and would tug on the rope three times if there was a problem. They gave me a flashlight, and I jumped in, lay prostrate, and just let the water wash me down.

As it got darker I used the torch and realized that it wasn't the boy we'd seen, just debris stuck there. Then I realized I'd actually got myself stuck, because the further I went down the drain, the more I backed the water up; I'd completely dammed myself in. I started to yank frantically on the rope, hoping that Gavin would understand. Fortunately he had realized what had happened

TO RESCUE A CHILD FROM A FLOODED DRAIN

and was in the pipe himself and yelling back to the surface to get every man and his dog to help pull me out.

It was a case of lying down and waiting for them to pull the rope back and trying to find a little air pocket every now and then to get a mouthful of breath before going back underwater. I was just hoping like hell that the rope would hold. It was terrifying. I honestly thought I was going to drown, but also going through your mind is: "Well, what if a kid is down here too? What's he going through?" You have a multitude of emotions and thoughts. It's pretty amazing actually because you feel quite matter-of-fact; you just think, "This is it, I'm going to drown."

I finally got to the surface and there were lots of police and SES. They said, "Well, what do you think?" I said, "Well, if he's down there, we'll be looking for a body, because there's no way he can survive that amount of water." Unfortunately at that stage his mom had arrived, but I'm not sure if she heard me say that or not. At the same time someone said they could hear a child screaming down the highway, so we jumped in the car and raced down there. There was traffic and people everywhere, and you could hear this boy screaming and screaming. Everyone was running around like chickens with their heads cut off. Some were suggesting we get a JCB (an excavator) to dig the pipes up.

We looked into the manhole and the water was bubbling up into it so you couldn't see down where you wanted to go. Somebody produced a ladder and Gavin went straight in and I went straight in after him. We got under the water and somehow popped up in a junction where six storm-water pipes met; a couple of them were flooded and just pouring water in on top of us, and somewhere in one of those pipes was this young boy. His screams were reverberating around the pipes, along with the roar of the water, and it was just the most unbelievable experience to know he was alive, but not know how we were going to find him and get him out.

We decided just to go up whatever pipe we could, but because we were going against the water it was terribly difficult. Gavin was crawling up his pipe and I went up mine, and I thought I could hear the boy in mine, so I went back and got Gavin. I was trying to crawl up this pipe and, because it was so slimy and muddy and there was water pushing up against me, trying to get a grip on anything was almost impossible.

The boy was still screaming out of control and I was trying to scream out to him, but it was just a waste of time. I was more or less dragging myself along with my fingernails and toenails and kept shining the flashlight ahead, and I think the boy must have seen some light, because he stopped screaming for just a short time.

I got close enough to see him, and I could tell that if he let go he'd come toward me, so I began screaming,

TO RESCUE A CHILD FROM A FLOODED DRAIN

"Let go, let go, come to me, come to me!" and he did. He let go and was washed straight into my arms, and I grabbed him. I said, "Young man, you just say thank you, God." He said, "Oh thank you, God." I pushed him, and the water washed him down to the ambulance guy.

Up at the surface people were yelling, "He's got him, he's got him." His mom was there, she'd heard him screaming for his life. News footage shows this half-drowned little boy being hauled up the manhole. The ambulance crews put him in a space blanket and his mom threw herself on top of him and the crowd just went berserk.

I got out and went back to the car, because I was physically and mentally exhausted. I was freezing cold, I just had my boxer shorts on, and to this day I don't know why I took my slacks off to go down there. One paper had an absolutely wonderful photo of me standing on the edge of the highway with my boxer shorts on—it was not a good look. I had blood and abrasions all over my back, so the guys took me back to the station for a shower, and I just sat on the floor and bawled my eyes out.

Physically, the boy was OK. He was very, very cold and in horrible shock, traumatized to the max. I'd been a police officer for more than twenty years. I'd seen death and carnage on a regular basis and endured a lot of trauma, but this was an eleven-year-old child, and his life was in the balance for so long; he had no way of knowing that help was coming. He suffered horrendously when he went back to school; he had kids and even teachers saying, "You should have drowned, you little bastard," because he wasn't a great kid. He continues to suffer, I believe. I organized for him and his mom to go for crisis counseling for a while.

About twelve months before this I'd been involved in an incident where two police officers were murdered, so I'd just been getting over that when this happened. I had a complete breakdown after the rescue, which was the end of my police career.

Some time later it was announced that I was going to receive the Cross of Valour, Australia's highest bravery award. For me personally it was extremely beneficial, because it gave me some sense that my actions were vindicated. I'd loved my police career and didn't want to leave.

I was saddened that Gavin didn't get the same award, because I felt he really deserved it. He got the Star of Courage, the second-highest award. We also each received a Galleghan award from the Royal Humane Society of New South Wales. At the end of the day it's a great story, because a child's life was saved—there can be no greater thing.

> *"My foot had basically rotted with gangrene."*

HOW IT FEELS TO LOSE A LEG

JOHN EDEN, *48*

I HAD A motorcycle accident in New Zealand when I was nineteen. I was riding home from work and going too fast on my bike, got into a slide, and hit a power pole. I was lying on the ground with my leg wrapped up my back and my head resting on my foot, and I thought, "Oh no, my foot's not supposed to be there."

I can't remember the next three weeks because I fell into a coma. Apparently I left this world a couple of times, but I can't remember any of it. They think I broke almost every bone in my right leg from my hip to my big toe. My foot was crushed. It was all a bit of a mess. It wasn't till I woke up and couldn't feel my legs that I realized I must have done something really bad.

HOW IT FEELS . . .

I was a rugby player and had just been picked for the junior All-Blacks. All I could think about was not being able to play rugby any more. Then I was worried about not being able to work—because I was a farmer.

I had fluid in my lower spine, which stopped the feeling going into my legs. The doctors were giving me lumbar punctures—sticking big needles in the small of my back to try and draw the fluid out. They did that two or three times and it wasn't working. It was very, very painful. The first two they did without anaesthetic and had to hold me down. They did it again after six months and the next morning I woke up and felt pain in my legs because the fluid had dissipated. I was hoping I could become normal again, be able to run around and do the things I used to do.

After eleven and a half months in that hospital they sent me to an orthopaedic hospital to have an operation on my foot. They put a bone through my ankle to relieve all the pressure. My leg was plastered up to my hip but I had these pains in my groin and the glands were up, so I went to the doctor and told him I thought my foot was infected. He said, "I don't think so, but we'll have a look." So they cut the plaster off and it was just all rotten. My foot had basically rotted with gangrene.

My leg was off within an hour because the poison could have gone all through my body. They took it off just beneath the knee and I went back to hospital for about another month or so. I was just lying there looking at the stump, and thinking how painful it was and wondering what I was going to do with my life now, wondering what the artificial leg would be like and all that. At that stage I got into drugs and alcohol because I was feeling really sorry for myself and couldn't play my sport, which I'd loved. I was always thinking, "I need another joint, I need another drink." I was heading in a bad direction.

The doctors sent me off to get an artificial leg made, and they just gave me a leg and some walking sticks and said, "There you go." They didn't even teach me how to walk. I got to the door, hung my walking sticks up and tried to walk from there to the taxi and fell over three times. I got in the taxi and went down to the pub and that was my life for the next eighteen months. I didn't handle it well at all.

One day a guy came into the pub and he had both legs off. He said, "You used to be a good sportsman; are you interested in trying out for the Paralympics in nine months?" and I said, "No, not really." He came back a few times and I just thought he was a pest. But I woke up one morning thinking, "I'm just not getting anywhere." I had a hangover and a shitty taste in my mouth, and it was rotten. So I rang this guy up and nine months later I won New Zealand's first ever Paralympics gold medal for an amputee.

I was a high jumper and did running and throwing. I started to realize that there was more to life than just drugs and alcohol and feeling sorry for myself. I went home and was a bit of a star—the papers and media were there and I was feeling pretty special.

In 1982 I came over to Australia because it has some of the best amputee runners in the world, and I thought it would be a good place to compete and to try to make a life for myself. I'd started playing rugby league in New Zealand, with just one leg, at senior reserve level. I tried to get into a Melbourne rugby club, but no one would take on a one-legged player. Finally Moorabbin gave me a shot in the fourth grade, their old men's team, but they complained that I played too hard for them, so they sent me up to third grade; then from third grade to second grade, and finally I started playing rugby for Moorabbin.

In 1984 I smashed my knee because I had no knee cap from the accident. I took my leg off after the game as I always did, to have a shower, and the stump just swelled up and never went down again, so they had to amputate it above the knee. That ended the rugby and the running. Not having that knee there slows you down incredibly, but this time I accepted it a lot better. I got the best artificial leg I could and just got on with my life. I took up discus throwing and went to the Paralympics in 1988 in Seoul, and I've been an international athlete ever since.

I'm very passionate about coaching rugby now that I

can't play at all. I spend a lot of time coaching athletics and rugby. I'm coaching nationally ranked athletes in the able-bodied field as well as the disabled field. I've been to five Paralympics and have won medals at each one except Sydney. Since Seoul I've been ranked one, two, or three in the world as a discus thrower—up to 2000, when I retired. I won the discus world championships in 1994.

Three days before I left, my fiancée, Mandy, got killed in a car accident. I wasn't with her when she died. The cops came knocking on my door to tell me that she had died, and I had to tell her parents. Losing Mandy was the hardest thing I had to go through, harder than all my accidents. I think losing someone close to you is much worse than losing a limb. That was a pretty traumatic time. There've been a lot of downs in my life, but there've also been a lot of ups, and I think tragedy makes you more tolerant.

I wish I hadn't lost my leg, but then I think, "Well, what would I be doing if I hadn't?" I'm pretty happy with my life at the moment. Since I lost my leg I've toured the world, and I think perhaps I'm more of an understanding, compassionate sort of person; I was probably a bit arrogant before. I didn't know anything about people with disabilities. Now I'm the national throws coach for the Paralympian team going to Athens and I do motivational talks and visit schools. Life's pretty good.

"I'd eat fast so I didn't have to share."

HOW IT FEELS TO WEIGH FIVE HUNDRED POUNDS

BOBBY BALLANTYNE, *42*

I'VE MADE BAD food choices all my life. I pretty much brought myself up, I fed myself and did whatever I wanted. Dad was always big on veggies and fish, which I hated, so I'd make myself three or four big, thickly sliced toasted sandwiches, then go to the store for ice cream, lollipops, or chips. I'd ditch Dad's healthy food in a hole out back.

I was always a very solid girl and I got bigger after I started having children. I've got seven kids. I got bigger and bigger with each pregnancy. At first, though I weighed a lot and looked huge, I could still do what I wanted to do. I wasn't happy with the way I looked, but I could still do things. But one day I couldn't climb on a chair to clean the top of the fridge, so I just didn't do it.

The heavier I got, the more things there were that I couldn't do, and it really started to get to me. I had bad depression. The worst time came when my weight went over 430 pounds. I couldn't eat when I was depressed or worried. I'd have to think happy thoughts to get in a tranquil frame of mind, just enough to be able to eat.

I never ate breakfast, but by late morning I'd have at least half-a-dozen pieces of toast with heaps of butter and grilled cheese; then I'd start on my block of chocolate. If there was cake in the house I'd start on that as well. I'd make myself a cup of tea, open a pack of biscuits, and drain the cup without putting my mouth to it—just one biscuit after the other dunked in the tea. I'd eat until the packet was gone.

I'd hide food from the kids because I didn't want to share it—it might have been a mud cake or whatever. I would cut off big pieces and shove them in my mouth as if I were in a race. I'd eat fast so I didn't have to share. Nighttime was my worst time for eating; I'd have my meal and then sit down to watch TV with whatever food I needed. When it was gone, I'd get one of the kids to get me more chips, lollipops, or chocolates.

It wasn't just the quantities I ate that were so bad, but the fact that the food was full of fat and sugar—the worst garbage you can find on the face of the earth.

I kept eating and getting bigger, and in my last pregnancy the scales couldn't weigh me, but I think I was probably about 380 pounds by the time I gave birth. The doctors told me I shouldn't have any more children, that it would be an absolute danger to my life. But I still didn't do anything.

There were nights when I would have really bad chest pains and I'd think I was going to die. I'd write letters to my children telling them how sorry I was, that I didn't mean to be like this. I'd make deals with God like: "If you let me wake up in the morning I'll do anything—I will go on a diet, I'll lose the weight." I'd gone to bed really scared, but I'd wake up the next morning and there'd be no chest pains, which made me happy, and I'd go and eat. I'd look at my kids and think, "I would do anything for you, anything," but I didn't do that one thing.

I could walk, but not far. If I had to check the letterbox I would be totally exhausted halfway to the gate. I thought, "That's it, Bobby, you're going to die, you're going to drop dead." I couldn't buy clothes. I had to wear ugly things just because they fitted me. I had a friend who made me skirts and tops that were basically tents. I chose horrible dark colors, because I thought they made me looked thinner. I couldn't wear underwear because there was none in my size. I bought super-size bike shorts and wore them as underwear to stop the chafing. I couldn't wear shoes because my feet and ankles were so swollen all the time—I had to wear sandals. I always felt totally ugly.

I slept in a chair in the lounge room for ten years because I couldn't lie down. I thought if I did I might die; the weight would all go up to my throat and I'd think I was going to choke. My husband bought me a recliner and I slept there. I tried many times to get back into a bed but it was impossible. I couldn't turn over, I couldn't move. I went through five chairs in ten years; I broke them all eventually.

I've been sleeping in a bed for the past year. I love going to bed—it's the best part of my day. It's one of the many magnificent feelings about losing weight.

I pretended not to be embarrassed about my weight. I became quite a rude person so that people would concentrate more on my rudeness than my weight. I was stared at all the time; friends would nudge friends and you could hear what they were saying. I was very embarrassed for my children; I knew what they must be coping with at school. People would talk to me loudly and slowly. The perception is that if you're big, you're stupid.

Chairs terrified me. They were my worst enemy—you could break them and end up on your rear, or you might not fit into them or, God forbid, they could have arms or be made of plastic. If I was invited out for lunch, I would get my son to check out the chairs, how far I had

HOW IT FEELS . . .

to walk, and if there were any stairs. If there were stairs I wouldn't go; unless the chair seats were solid wood without arms I wouldn't go. It was a life of restrictions.

About eighteen months before I started to do something about my weight, I was diagnosed with type-two diabetes; it didn't mean anything to me though. I probably spent the next year eating more chocolate and sweets than I ever had. I did start to think more about it though, and it began to scare me.

The big crunch came one day when I was having a shower and I slipped. It was a feat to get in there in the first place. I was really, really scared—there was no way I could get up. I called for my son; he came in, took one look, and started crying. I said, "The fire department will have to come and take out the wall." He said, "Please, Mom, just try to get up." So much was going through my mind, like: "How can you do this to your children? This beautiful young man just destroyed." I was scared for the kids; that it was going to be on the news.

I sent my son to get every pillow he could find and we lined the bath with them. He had to get a pillow under me but the sheer strength I needed to raise my own weight was huge—even for one second it was torture. Finally he did it, and I was laughing and crying.

Then I had to get onto my knees; I can't really describe the pain of that—I've never had a big truck sitting on top of me, but that's kind of how it felt, it was absolute agony. But I did get onto my knees. Grabbing hold of the sliding door, I got there. I was crying harder and harder. My son ran and got a kitchen chair and I used it to lever myself up. He sat on the chair and I had to lift five hundred pounds. By some miracle I did it. It was one mighty heave. My son was cheering.

And that was it for me. It wasn't the agony of getting up or the risk of the fire department—it was my son and my daughters. My daughter and I joined Weight Watchers, but they couldn't weigh me because their scales only went up to 420 pounds. I had to be weighed in a back room on old-fashioned scales.

What depressed me was being told I had to lose 340 pounds, the weight of several entire people. I started to worry that I couldn't do it. Then they told me that when I lost eleven pounds I'd get a bookmark; it was my first goal. I wanted that bookmark. I did my first week, and pretty much from the start I got rid of the bad foods.

I lost almost eleven pounds in the first week; that spurred me on. I lost eight the second week. I just kept setting little goals for myself. My biggest goal was to get off the old-style scales and be weighed like a normal person. In the past eighteen months, I've lost 295 pounds and counting. My goal weight is 160 pounds, and it's well and truly within sight.

> *"I had blood coming from my eyes, nose, and mouth."*

HOW IT FEELS TO CHOKE TO DEATH ON A CHEESEBURGER

Cameron Lewis, *18*

ME AND ALL my friends had been out doing a bit of late-night shopping as we always do on a Thursday night. We usually stop off at McDonald's or somewhere like that just to grab something to eat on the way home. Stuffing a whole cheeseburger in my mouth is a party trick that I had done before, so this wasn't the first time. My friend Chris said, "Show the other guys, show 'em, show 'em." I was a bit known for it. So I went inside and ordered a cheeseburger, came back out, and put the whole thing in my mouth.

But then I got the giggles and I sucked the whole thing straight back into my throat. It got stuck in my windpipe and that's kind of where it all went wrong.

TO CHOKE TO DEATH ON A CHEESEBURGER

Because I'm always a bit of a joker— a bit of a knucklehead—my friends just thought I was fooling around when I first started running around the place pointing at my throat. After a couple of seconds, when I started to go blue, they caught on and realized, "Oh, he's not playing around."

I started to panic, I couldn't cough it up. I was trying and trying and all my friends were trying that thing called the Heimlich maneuver—and bashing me on the back. Then they all started to panic as well, and a couple of them called the ambulance. I was just choking and going purple. I had blood coming from my eyes, nose, and mouth, because I couldn't breathe and there was blood coming from everywhere.

I was going absolutely crazy running all over the place. My friends were shouting and carrying on. It was awful, I was just terrified. I was thinking, "This can't be it, I'm too young, I don't want to die, I don't want to die," and all that kind of stuff. It was pretty full on. It only hurt a little bit, but I was starting to go a bit numb because I was in shock. After about five minutes without air, I just blacked out. I was on the ground trying to breathe and everything just went black. It was like falling asleep and dreaming.

It only took the ambulance about a minute to get there. Apparently a couple of the ambulance guys had a go at banging me between the shoulder blades, but it was no good. They had to get these forceps down me throat and drag as much of it out as they could. Because the bun was so soft, the saliva melted it and just made it like glue. Then they had to force a tube down my throat to try to get some air into me. My friends all thought that was it for me. They thought I was gone.

Then they took me to hospital. It happened on the Thursday night and I woke around Saturday lunchtime in intensive care with a tube down my throat. I was on life support and in a coma for about twenty-four hours. They think I was without oxygen for eight to ten minutes. They say you can go completely brain dead after three minutes, so I was pretty lucky.

Dad lives in Brisbane, and my mom and I live in Toowoomba, but that day mom was down the Sunshine Coast on a business conference. The hospital called dad and told him that they didn't think I'd make it through the night. So he called mom and said, "Cam's in a bad way, they don't think he's going to make it till morning, so you'd better get up here." So her company paid like $250 for her to get in a cab. I was so close to dying that the hospital talked to mom and dad about whether they would donate my organs. Nobody left the hospital for about thirty-six hours. All my family were expecting the worst.

When I woke up the worst thing was they'd stripped me of all my clothes, because they had to put wires and tubes everywhere. So I woke up and went, "Okay, this isn't good." I wasn't happy about being naked. It took a few seconds before I remembered what had happened. The first thing I asked for was a notepad and pen, because I still had the tube down my throat, and I just wrote to everybody, "I'm sorry."

In the hospital, once they were confident that I could breathe on my own, they took the tube out and it was fine after that. Once I woke up from the coma I just got better in leaps and bounds. But I still had bits and pieces of cheeseburger stuck in my lungs that I had to cough up. That was just so gross when they said I had to bring it all up again. The doctors told me I was just the luckiest person and that I should buy a lotto ticket. They were amazed that I got out of it without being brain dead. Ten minutes without oxygen is unbelievable. It was pure luck. I've got no aftereffects at all.

While I was in intensive care, my friends weren't allowed to see me because it's family only but, once I got transferred to the regular ward, they came in. They all felt really guilty because they hadn't been able to do anything to help me. I just reassured them that there was nothing that they could do—I mean, the trained professionals couldn't even get it out. All my friends and I, we watch out for each other, so it was pretty tough for them.

All in all, I had five days in intensive care to make sure I wasn't going to slip backward, and then, after another two days, I went home. I don't take anything for granted now. I mean, I should have been dead; they don't know why I lived and how I managed not to get brain damage, so I just enjoy time with my friends and my family. I still do teenage stuff, fooling around and that, but every day is a blessing to me. I've actually joined the St. John Ambulance now. It's something a bit different, I guess, and hopefully I'll be able to help someone else. It took me about nine months, but I've started eating McDonald's again now, but not as much as before!

> *"On one ski I wasn't traveling fast enough to outrun it."*

HOW IT FEELS TO BE
CAUGHT IN AN AVALANCHE

Mike Balfe, *44*

I WENT TO Alaska with a group of friends who were all serious skiers. On this particular day it was bright and sunny and hadn't snowed for about three weeks. The guides who were leading us were searching for untracked snow, because that's the whole reason you're up there. You go to Alaska basically to experience powder skiing.

In Alaska you only ever ski one at a time, and the guide usually goes first. We arrived at the top of the slope that our guy had chosen for us, and he was a little bit suspicious. He ended up digging a pit of some description to test the stability of the snow. I'd never seen this done before. He felt that there was something not quite right, so he dug this pit, which would have been about four feet deep, and then he carved out a number of blocks of snow to check whether there were any sheer layers near the top. He did a test where he tapped ice crystals onto a mirror to study their shape. He decided it looked OK; if he thought it wasn't right, he would have called the helicopter in and we would have been lifted to another slope.

In Alaska the slope is very steep, and it then either plateaus out or ends up on a glacier, but where the slope meets the glacier there is often a deep ice crevasse. We were to ski down about eight hundred feet and end up on a glacier. You don't go to Alaska unless you're an extremely competent skier. You're really out there, and accidents do happen.

Anyway, the guide went down, and after about two turns he vanished. We couldn't see him any longer because the slope was so steep. He came out onto the glacier and radioed back up to say, "The skiing is fine; conditions are good." I skied down and chose to go basically where the guide had gone, so for the first four or five turns I was looking to line up where I was going to cross the crevasse over a snow bridge. Then I started into the run proper, and when I was halfway through a turn, I saw in my peripheral vision that the entire slope I was on had given way.

I could see a crack forming and then it opened up behind me—probably only about fifty feet behind me.

TO BE CAUGHT IN AN AVALANCHE

The entire slope moved, and I was fifty feet into it, so there was another 650 feet below me that was moving—that was the good bit, but I wasn't thinking that at the time. I shat myself. I was very fearful and believed my life was in danger.

As I pressed back down into the snow, my downhill ski was removed by the movement of the snow, so I was on one leg. I tried to go straight to get out of the area, but there was no way that was going to happen—on one ski I wasn't travelling fast enough to outrun it. I got knocked off my feet, and it surprised me how lucid you can be at that precise moment when you know you are in danger. My life didn't flash before my eyes, as the cliché says it does, but I did think of my family—that was the first thing that popped into my mind.

Then the thing I was concerned about was the crevasse at the bottom of the slope. I attempted to cartwheel and stay above the snow, so I didn't get sucked in and get buried. I succeeded in staying on top of the snow and somersaulted, trying to stay as tall as I could. I was all arms and legs, trying to hand-spring—not that I'm an acrobat or anything. I don't know how many times I flipped over—it could have been fifteen, it could have been twenty-five; I know it was a lot.

So I'm plunging down this hill and my only thought is to avoid going straight into the crevasse, that I should try to flip over the crevasse—not that I had any concept of where I was, because I think I shut my eyes. The guide was the only one who could see this happening, and according to him I did a backflip over the crevasse and landed quite heavily on my back. That took most of the momentum out of my fall, and I think I flipped over twice more and ended up on my feet, facing up the hill. My equipment was all over the place; it was just me standing there with my backpack on and feeling very thankful.

I was in shock; I was shaking like a leaf. The guide was yelling at me to ask if I was OK, and I put my arm up and waved because I couldn't talk, I was so winded.

The guide had radioed the helicopter and it was on its way. They put me in it and sent me home to the lodge. I was extremely bruised and battered but that was all. I was very lucky not to have any other injuries. We took the next day off, but I skied for three days after that.

Now when I think back on it, it still disturbs me. I'd never given avalanches any credence, so to speak. I'd never worried about them. I've skied in many, many places that potentially could have been very dangerous but I never really worried. I always thought I was good enough to ski out of it or get away from it. Well, I don't think that way any more, and I will be a lot more circumspect going into untracked terrain.

> *"He had no pulse, no nothing — he was just dead."*

HOW IT FEELS TO SAVE A LIFE ON YOUR WEDDING DAY

JODIANN, *38*

SIMON AND I had known each other for only a few months when he proposed to me. The first time we met it was only for about twenty minutes; then it was three months before I saw him again. He turned up on a motorbike; I waited all day for him. I walked up and planted a kiss on his lips and we were engaged two days later—it was very quick. Less than six months later we got married.

I had only recently moved to Perth after working in Kalgoorlie as an emergency nurse. We had both always wanted a really romantic, beautiful wedding. Our wedding cost a fortune; we planned a huge thing. We had an enormous marquee set up on a local peninsula, and we were to go by boat from the garden where we got married, up the river to where the reception was going to be held under a huge tree covered with fairy lights.

It was absolutely beautiful.

I had a Scarlet O'Hara - style dress with a heavily beaded bodice, very tight waist, massive hoops, and a really long train. The ceremony went well, but the second I got in the car to go to the boat I said, "Let's skip the boat and go straight to the reception." I had a definite feeling that something wasn't right. But Simon said, "We can't do that. John has been waiting for ages to boat the bridal party to the reception." He was our best friend's father and was thrilled to have a role in our wedding. He was sixty and a dear friend. I felt strange and said again, "I don't want to go to the boat," but it was already organized.

We got on the boat and set off. There were two bridesmaids and two groomsmen, as well as John's wife, son, and daughter. We had only just got underway. I had snapped a suspender at the back of my leg and one of the bridesmaids was literally up my skirt trying to fix it; my leg was up in the air. Everyone else was getting their champagne organized.

I was the only one looking at John, the skipper. All of a sudden I saw that he went very gray, and with that he just salivated really badly and collapsed onto the throttle lever so that the boat went flat out, head-first into the canal wall, and sent us all flying. My bridesmaid was jolted up my body because she was still under my skirt; she ended up almost in my bra. We all fell backward as the boat jammed into the wall. Everyone screamed. John collapsed. He had no pulse, no nothing—he was just dead.

I started resuscitating him. One of the groomsmen grabbed me and said, "You don't get involved. It's your wedding day!" I said, "I'm the accident and emergency nurse. I have to get involved!" All the guys picked John up and laid him flat in front of me. They had to move stuff out of the way so he could lie flat on the bottom of the boat. I said, "I need this here, and this moved," and everyone just did it.

There was chaos initially and then everyone was calm and they sat down. All I could think was, "I have to save this guy." I picked up the hoops of my wedding dress and straddled him to do mouth-to-mouth. His daughter dove overboard in her lovely wedding frock to swim up the canal to get help. His son grabbed the controls of the boat and pulled us away from the wall and kept the boat still so I could work on his dad.

All of a sudden John's heart just started beating by itself. I was about to start heart compressions when he slowly came back. He hardly had any pulse, but then it got stronger and then he came back to consciousness. It was just amazing. It can only have been divine intervention. I turned him on his side and his color came back. I said to him, "Do you know where you are?" He said, "Wedding day," and he was crying.

His daughter had swum to shore, told two other guests what happened, and they called for the ambulance. At the shore the ambulance workers came onto the boat and took him to the hospital. They were a bit surprised to see a bride resuscitating him. It turned out he had a heart blockage. He had a pacemaker put in that night.

We then drove back to a friend's house where I went to fix up my lipstick—and it hadn't budged. It was a Poppy lipstick. Ever since, I've been meaning to write to Poppy and tell her. I wiped it off and applied it again anyway, after all that breathing on John's lips. We finally got to the reception an hour-and-a-half late, but it was still a good day, because he lived. I think a wedding's a blur for everybody. The only part that was ruined was the pre-drinks before the reception started.

When John collapsed, I remember thinking, "What the . . .? Do you mind? On my wedding day?" But it felt good to save someone's life. When he came out of the hospital they were playing "Here Comes the Bride" on the radio, and he cried and said he was the first person to kiss the bride and that had made him happy. John was incredibly thankful. His son came to see me the next morning and wrapped his arms around me and said, "You're the best," and the love in his heart was amazing;

that alone made it all worthwhile. John later gave us the most beautiful cheval mirror that he had made for us.

The annoying thing was that my wedding dress survived everything, with me crawling around the bottom of the boat and so on, but then when we got out of the taxi at the end of the night it got grease on it. I later realized that the catering staff at the reception had taken my horseshoes—stolen all of them. But when I look back on my wedding day I still think it was a beautiful day.

> *"I was to go back and help people overcome the fear of death."*

HOW IT FEELS TO DIE AND COME BACK

Ken Mullens, 56

I WAS IN the hospital following a heart attack. I was having lunch and I felt very strange. I had a cardiac arrest, and then it was instant death, but I felt I was aware the whole time. It was a strange, strange feeling. It was like when you wake from sleep, but it really wasn't sleep; then next I went into darkness—really, really black. There was not one speck of light. I used to suffer from claustrophobia, but this darkness didn't concern me. It was peaceful, very, very peaceful, and then I felt movement as if I were going somewhere, but I didn't know where.

I was drifting, but I wasn't sitting up or standing up. I was just drifting in the form that I was in, whatever form it might have been. At what speed I don't really know, but I know it was incredibly fast, as fast as electricity, you might say. That speed factor was an important point; it was very, very fast. Eventually, after a period—I can't say how long, because there was no sense of time—the darkness dispersed slowly, ever so slowly, and then a light appeared and it seemed as if it were a million miles away.

It was like a hundred thousand suns. Bright, incredibly bright. I could look directly into that light. It was so very powerful and ever so bright. Words can't really describe the magnitude of the all-consuming love experienced when in the light. And not only love, but perfection, peace, serenity, calmness, and beauty. I felt that I was safely home. I was overawed by the experience.

The wisps of cloud are the things I can still remember most clearly—I just put my arms out to feel those clouds and I couldn't feel anything. Then I looked and I had no arms, and then I looked down and I had no body. Strange as it must seem, I had 360-degree vision, because I looked behind me and saw I had no back. I did fully comprehend the fact that I was probably the shape of a ball, but it didn't frighten me—it was a feeling of "oh well, so what."

I felt that there was a form in the light that I could see, but it was only vague. I felt ever so humble. I felt that this was something greater than great and I shouldn't have been there. I couldn't see much more than what I imagined would be a head, but there were arms, or what I thought were arms, in the light. They were outstretched and they raised me up. I felt ecstatic, but I still couldn't see into the form. I could look into the light, but I couldn't define that figure.

I knew what I had to do. I can't say words were spoken because there was no such thing as speaking—it was like mind language, and my mission was made very clear. I was to go back and help people overcome the fear of death. I would go back and write two and a half books. I've written two, but I get apprehensive about the third!

The experience was just so powerful, so all-consuming, that the difficulty I had as a free spirit was fitting back into the human physical form. I just didn't seem to fit—I'm not talking size; it just didn't seem right. I'd gone back home, back where I came from, and then I had to return here and fulfil this mission—help people by telling them that in death, your mind, soul, spirit never cease.

Later I could tell people what had happened in the hospital room while I was dead. It was not like I was sitting on the ceiling, because I was traveling towards the light, but somehow I could also see what was happening there; I knew exactly what was being done, where and when—it was like being omnipresent.

TO DIE AND COME BACK

The doctor said I was dead for twenty minutes; they'd stopped working on me and started packing up, and then all of a sudden, they said, I gave a cough and I was back. Afterward I went and spoke to all the nurses and doctors about it. They said that when I revived I was still black and purple. It was the longest time they'd had someone gone after a cardiac arrest before being revived.

I think I had to die to learn to live. It changed my whole outlook on life, my whole outlook on people. It changed me from a bigoted type of person to someone more broadminded who can let people accept what they want to accept. As a Christian I used to feel my faith was the only faith, but this experience made that view seem ludicrous. I firmly believe that people fear death because of the uncertainty of how and when they will die, but I now know that death is just going into another chapter, another phase of your existence.

> *"We were broke — more than broke."*

HOW IT FEELS TO
WIN THE LOTTERY

Roger, *53*

I WAS A sergeant chef in the army in 1986. I came home one night and heard the draw on TV. My wife Kaye threw her pen in the air and called out, "We've won it!" I had to check the ticket several times just to make sure. We didn't have any money to celebrate that night, so we borrowed money to buy a carton of beer. I rang my mother straight away to tell her.

There wasn't much sleep that weekend; we were very restless. We didn't know what to do or what to think. There could have been heaps of winners or we could have it all to ourselves. It was a long weekend, so we had to wait from the Saturday night until Tuesday morning to find out what we had won. In those days you didn't register or anything, so we had to hold the ticket in our house. It was hidden down the bra, under the pillows—there were about ten different places.

We thought we might have won enough to buy a house—about fifty or sixty grand. We were broke—more than broke—at that stage. We were paying off a car, paying off furniture loans, like everybody else. I was in the army, hoping to stay there long enough to be able to buy our first house, and Kaye worked too. Kaye always bought lottery tickets, she'd been doing it forever, but it was the first Quick Pick she'd ever bought. Some woman that Kaye knew pushed in front of her in the line when she went to buy the ticket and Kaye just let her go—and that meant we got the winning ticket.

At two o'clock on Tuesday, I was in the car when I heard on the radio that there had been only one winner. I knew then that we had scooped the pool. We won $1.4 million. I was by myself in the car. I had to pull over and have a smoke, my hands were shaking that much. It's something that you don't react to, even now you just can't comprehend it.

We took the ticket to the bank and they controlled everything. We never actually saw any cash; all we saw was the bank account statement showing that the money had gone into our account. I decided to get out of the army. I was only staying in to do my twenty years of service.

HOW IT FEELS . . .

The first thing we did was pay off all the loans that we owed. We bought a house, and a new car each. We'd never owned a new car in our lives. We went to Singapore for a holiday, which was a real big step for us—it was the greatest experience we'll ever have in our life. We'd never been overseas before; we could never afford anything like that. I bought myself some nice fishing gear and Kaye refurnished the house. We had great pleasure in giving our other furniture to a friend who was really broke—she's still got it, I think.

But nothing was like we thought it would be when we won. You think it's going to be the best thing ever, but it's not. The best thing about winning was being able to give money to our family. That was fantastic. It was actually better than we thought it would be. We gave away $20,000 for each brother and sister and our parents—$260,000 in total.

When I gave one sister her share she started crying and said, "You know what this means to me? This means I'll never have to work again." That meant the world to me. She went and told her employers she was quitting.

Winning the money hasn't meant that Kaye and I have had no ups and downs, but I think it's probably brought us closer together. We're probably old fashioned, but we don't want anything in life except to have a decent little house and just be able to afford to do things. But that money doesn't last forever. Now we're comfortable but we're not what you'd call rich people.

The worst thing about winning was the way people treated us. In a small town like ours, everybody knows you. Some people say that you change when you win money—but it's not you, it's the people around you. We got heaps of letters from people asking for money. People who I thought were our friends changed their attitude toward us. I got into a couple of terrific bloody fights in the pub, so I stopped going. People would make smart-ass remarks about how I could buy a round of drinks for the entire bar—just jealousy. In the end they drove us out of the town.

That's the thing that really pisses you off about it, the way that some people treat you as though you're different. I know for a fact that I've never changed; it's never changed me any. A couple of times since winning I've been walking up the street and thought, "I've won $1.4 million dollars," and it doesn't mean a thing to me. All it means to me is that I don't have to go to work and answer to somebody. I haven't worked in a regular job since we won, but in the past fifteen years I've cut and sold wood every year to earn a small wage to pay bills and do what normal people do.

After twenty years of having the money, and a lot of bad investments, thanks to people who were so enthusiastic to give us their "expert" advice, things just went downhill. We lost $70,000 overnight when the stock market crashed. Kaye was always very good with money. If I want any I just go and ask her for it. I'm a fool for getting caught out down the street with no money. I never walked around with a couple of grand in my pocket. With our home and investments we wouldn't have a quarter of the money left now.

I'm into the simple things in life—a boat so I can go fishing when I want to, that's all. But you know, life's struggle never ever stopped. The struggle is still there. Our son recently had a car accident with his two little girls—they were all fine, luckily. Our daughter is driving up and down the highway all the time and the worry about that never stops.

People think when you win money you're immune to all the shit, but you're not. You still get it all. It's been a real battle with our two kids; they're like everybody else's kids. Money can't buy you health or safety. You worry about your kids all the time, and now that they've got their own kids, you worry about them, too.

> *"You have enough time to know what's happening and to know it's the end."*

HOW IT FEELS TO BE *CRUSHED IN AN ICE CREVASSE*

Raina Plowright, *29*

I WAS GOING to work on a penguin-monitoring program at Mawson Station in the Antarctic, but I was only there for three days before the accident. We were being trained to use the quad bikes, but the person leading the training accidentally took us through a crevasse field. The ice was very, very bumpy and very difficult to navigate, so I was really concentrating.

I saw a little patch of snow in front of me and thought I'd better be careful about how I got across it. I slowed down and suddenly my front wheels were just spinning on the ice, and then my back wheels broke through the snow—it was a snow bridge over a crevasse—and I found myself plummeting down. My body was upright and the four-wheel bike was plummeting vertically with me.

HOW IT FEELS...

Even though I must have been falling for just a millisecond, I had enough time to think, "I'm falling—this is it."

I always thought when someone fell off a cliff it would happen so fast they wouldn't know anything before they were dead—which I thought was good, because they wouldn't have that horrific, awful feeling of knowing they were dying. But it's not true; you have enough time to know what's happening and to know it's the end.

The crevasse was at least a hundred feet deep. It was very dark. The ice was a neon blue color. After falling about twenty feet, the next thing I felt was just this enormous weight of metal crushing me as the seven-hundred-pound bike fell on me and wedged me into the ice. It was crushing my thoracic area and my abdomen, so I couldn't breathe at all.

I thought, "Well, this is definitely it now," because I couldn't get a breath. I thought I was going to die within minutes. There is not a word to describe that feeling. It was extraordinarily unexpected—to be riding along on the ice and then two seconds later I'm dying. Here I am, halfway down this crevasse, not able to breathe, and I thought, "What a stupid way to die."

I totally focused on figuring out some way to breathe. My chest was crushed, my abdomen was crushed, I had no mechanism to move air into my lungs. I had done a lot of meditation in the past and I don't know if it saved my life or not, but I pulled all those skills together to try to

TO BE CRUSHED IN AN ICE CREVASSE

breathe, and I was able to move some air into my lungs through my throat. I'm not sure exactly how I did it.

The people above came to the edge of the crevasse and were yelling down. I couldn't communicate back, but one of them asked me to wiggle my fingers and I knew it was really important to show them that I was alive. So I did it, but I lost control and couldn't breathe again and had to regain focus.

Very soon after being crushed into the ice wall I felt my legs go numb. It was like I had two heavy logs hanging from my body, and I also lost control of my bladder. I'm a veterinarian, so I knew instantly that I might be a paraplegic. That wasn't scary for me at all, though. At the time I just wanted to live, and if I was a living paraplegic that was just fine.

When the search-and-rescue guy broke the ice above the crevasse and came down to me, the first thing he did was to check my pulse and I just thought that was such a waste of time. I was on the edge between life and death and here he was checking my pulse! I was thinking, "Just get me out of here."

He put a rope around my chest and then attached me to his harness. They put a winch on the quad bike and started to pull it up. But the movement crushed me even more into the ice wall; I was being compressed into it. They pulled it up faster and then I got a real breath and fell into the guy's arms.

HOW IT FEELS . . .

A white curtain of complete despair came over me and I lost consciousness. As I was lifted over the edge of the crevasse, I regained consciousness, and I felt this overwhelming feeling of "I'm OK, I'm going to live." I was down there for an hour and a quarter but I can only remember about three minutes of it.

They took me on the back of a truck to the medical center. My body temperature was about 90 degrees F, (the average body temperature is 98.6° F), but it was fifteen minutes before I started to feel cold. I was shaking violently and felt extraordinarily icy to the core. I heard the doctor say, "Her blood pressure is sixty over zero," and I thought that just couldn't be right.

I had overwhelming nausea and started vomiting blood. I just felt so awful; the pain was extraordinary. The next day they decided to operate. I had a crushing injury to the abdomen, bleeding vessels, a haematoma in my abdomen which was crushing nerves in my spinal cord, and traumatic pancreatitis.

Both my legs were paralysed but one started to come back straightaway and the other recovered in a couple of months. I had a second operation a week later and then they sent me home by ship. I was in a rehabilitation hospital for three weeks back in Australia. I don't think your body is ever the same after injuries such as those, but I'm largely recovered.

> *"The dog head-butted me in the face, bit me on the breast, [and] slammed me against the side of the truck."*

HOW IT FEELS TO BE MAULED BY A ROTTWEILER

MARY, *Age withheld*

I WAS ON my way to work as normal. I was standing on the side of the road in line with the outer edge of parked cars, just watching the oncoming traffic. I noticed one of those one-ton-tray-trucks (pickup truck) coming towards me. I saw that it was fairly close to the parked cars and that there was a big, black rottweiler hanging out the side. The next thing I knew, Wham! My world changed a little, my plans changed a little—to say the least!

The dog head-butted me in the face, bit me on the breast, slammed me against the side of the truck, which broke my arm in several places, and then dragged me along for about twelve feet until I fell and landed in the road. I was so lucky that the driver in the car behind the truck was paying attention, otherwise I could have been run over as well.

There was a crowd of people around me very quickly. They were yelling, and I was looking at the asphalt and all I could see was the underbelly of cars whizzing past me. I did a quick whip around of my body as I was lying there, checking to see if I was alive, what's broken, what's working.

People were trying to get me off the road but I kept saying, "No," because it hurt so much when they tried to move me. I heard a fairly aggressive male voice having a go at the dog's owner saying, "Why have you got a vicious dog like that?" At that stage I thought the dog had bitten me because I got such a whack in the face.

I was really, really lucky with where it happened, because I was right outside a large building site which had trained aid officers. They heard the bang and thought I'd been hit by a car. They came out and put a witches' hat up to block off the four lanes of traffic. I so easily could have been killed by a passing car.

You could tell I was a caffeine addict because I still had the two dollars in my hand for my coffee before work, and I was thinking, "Oh, my coffee." But really I had much more serious things to think about! Because my work was just across the road, I asked one of the building guys to let my boss know that I wouldn't be in to work that day.

When I was being wheeled into the ambulance, my

boss and the woman from work who handled worker's compensation were in the crowd and she was yelling: "Don't worry, its worker's comp!" That was fabulous news, because all you can think is: "This will cost a bomb."

I was in a lot of pain at that stage. I didn't realize that my breast had been bitten, but I could see some blood on my T-shirt. Later that day, when I was in hospital, the police came to take some photos of the wound and I saw that it was really meaty and bloody around the nipple. I found out that there had been a bite about the size of an American half dollar next to my nipple—I joked that I got a nipple piercing that I didn't anticipate.

The worst thing about the wound was that it went quite a few centimeters deep so they had to do a lot of reconstruction in surgery. I had about twenty stitches on the inside and the same amount on the outside. I also had large teeth marks around my breast, and some of those punctures needed stitches as well. There was also a lot of bruising, which came out the next day. My arm was dislocated and several bones were broken. I have a plate and some pins in there now.

The attitude of my workplace really helped in my recovery. I think it would have been very different if I didn't have their support. I was in hospital for several days and had surgery to clean out the wound and prevent infection.

> *"My situation is life and death . . .
> but you can't see the wound."*

HOW IT FEELS TO BE SUICIDAL

Grace, *35*

I ALWAYS THOUGHT suicide was such a dramatic, theatrical act. A tortured soul flinging itself into eternity. Now I know better. I know that suicide is not dramatic, there's no excitement or morbid thrill. When you feel suicidal, it's literally the last gasp of life, when you can no longer summon the energy or will to continue living.

There's no chilling crescendo of music, as old movies would suggest; it's all dull and dark and hopeless. That's how it was for me, and how it may be again tomorrow—I don't know. I hope the feelings don't come back, but as I fight my way back up to the surface through the stifling layers of depression, I've come to realize that the urge to die is insidious. It is just there sometimes.

TO BE SUICIDAL

At my worst I stood—a thirty-something happily married mother of young children with a successful career—contemplating a handful of pills. They looked so pretty, so inviting. I knew if I swallowed them the pain would stop and that's all I cared about. Looking over my husband's shoulder one day as he flicked through a hardware catalog, I noticed the razor blades. They sparkled so attractively, looked so efficient and clean, and I wanted them, wanted to feel them open up my skin and let the sullied blood flow away and take me with it.

I'd talk about these suicidal thoughts with people I could trust, and without fail they all said, "You're so lucky, your husband is gorgeous, your children are gorgeous, you have so much to live for." I know they meant well, as if I hadn't already tried all that positive thinking. That's the point—depression robs you of the capacity for positive thoughts.

My husband cut a tendon in his finger once and we took him to the local ER, where the nurse asked him to try to bend it. He couldn't and almost fainted when he realized that he had no control over the finger, that his brain was sending commands, but the connections were gone. That's how being suicidal feels to me. My brain logically knows all the positive stuff but somehow the message doesn't get through. If the connection were working properly, I wouldn't be suicidal. If I'd said to my husband, "If you just think positive your finger will be OK," the hospital staff would have laughed at me, yet every week I'm encouraged to think positive, not give up, count my blessings, and look on the bright side. I do, I do, I do—but my illness is as real as his cut finger and, with all the will in the world, I can't make it better by putting on a happy face.

Sometimes I get so terrified because I feel that it's inevitable that I will die from this . . . this pain. That in the end I will have no control or choice anyway; my body will take the only option it has for ending the hurt. I don't want to die. I have never wanted to die.

As I see it there are two choices: to live with the pain or to stop the pain. I'd like it to stop—it's unbearable, never-ending agony. Although I have cut myself, have counted the pills to make sure I have enough, have thought about where on my chest a gun should point to ensure a direct hit in the heart, and have contemplated buying a hose for the car exhaust, I'm still here.

I'm working through the depression and the shocking childhood trauma that triggered it. I'm doing all the right stuff: seeing a psychiatrist, taking the prescribed medication, and talking, talking, talking. I'm trying so hard to be well. I'm devoted to my family and want to be well for them, and for me. I'm working so hard, but sometimes the hardest part is that the illness and the work are invisible. I see moms at school rally around a parent who has had surgery—with casseroles, visits, and flowers—and I feel isolated and alone. My situation is life and death too, but you can't see the wound.

I chat and smile and try to appear normal when I can, so outwardly I can seem fine. They don't know that I go to bed for days at a time, that I have dreadful nightmares, can no longer seem to work, though once I was a workaholic, have taken to watching daytime TV, and often feel so terrified and overwhelmed I'm paralyzed. They don't know that I sit in the small hours of the morning reminding myself of the damage I would do to my children if I took the easier option. I hate myself most of the time. I feel empty, shattered inside, and deeply, endlessly sad for the childhood I lost.

I don't know how my story will end. I'm smart enough to let good friends and family know how unwell I am, and I promise to call them whenever the weather in my head turns ugly again. I want to reach out for help; I don't want to be alone in the swirling ocean of my mind, where the water is frigid, dark, and bottomless, the sky thunderous and dark, and I'm cast adrift, desperately treading water—with no land in sight. I want to get better, I want to be the fun-loving, motivated, involved mom, sister, wife, daughter, and friend that I used to be. I want to achieve at work again, cook muffins, and have picnics in the sun with the kids.

I want to put all the pieces of my shattered soul back

together again—it will be precarious because there are shards of broken glass inside me and they need to be handled with care, for they are jagged and will wound. I feel as if my soul is a mirror smashed by pain. These pieces jangle discordantly inside me and I have to move through life carefully so I don't get stabbed by the edges. But if I can just get this broken mirror back together, maybe one day I'll look into it and see, reflected, me—smiling.

> *"It felt like the biggest whack to the side of the head that I've ever felt."*

HOW IT FEELS TO BE SHOT IN THE HEAD

SENIOR CONSTABLE NEIL SAUNDERSON, *35*

I WAS ON night shift in 1992. It was just a routine shift on a quiet night. I'd been in the police force for less than a year at that stage. At about 1:00 A.M., myself and the guy I was with were just driving along Beach Road and we saw two guys running down the road. One of them was holding what looked like a briefcase.

We decided to see what they were up to, so we did a U-turn and shone our searchlight onto the front of the houses on Beach Road, but we couldn't see anyone. Then we did the same on the beach and we still couldn't see anyone, so we did another U-turn and parked the car out in front of the houses.

I was in the passenger side and I jumped out and

walked in one driveway, and my partner walked in the next one. I shone my flashlight up the side of the house, which had a big fence. I looked behind the fence and there was nothing there, either. There was nothing up the side of the house.

I walked a bit further into the front yard and shone my flashlight onto the other side of the house and there was this figure of a man with a gun. The next thing I remember is the muzzle flashing—getting shot. I sort of put my arms up but it was too late, and all I saw was the flash. It felt like the biggest whack to the side of the head that I've ever felt, and I wondered what had happened.

I came staggering out of the yard and my partner thought that I'd shot someone, not thinking that someone else would have a gun. He came flying around and saw me with blood streaming from my head and said, "What's going on?" I said, "I think I've been shot." We didn't have portable radios on so we had to try to wake up the neighbors in the next house, but there was no one there. My partner jumped the fence to the house next to it, and I sat on the doorstep of the neighbors' house.

The bullet hit me just above my eye, and broke the bone around my eye socket. It just missed my eye. It went towards my temple between my skin and my scalp; a little fragment of bone came out at the back of my ear. The guy had a sawed-off gun. Earlier that night, he'd threatened someone else with it. He was like a powder keg, and I just happened to be in the wrong place at the wrong time.

Being shot was really painful. I've been hit in the head before, playing soccer or whatever, and it was like that but ten times worse. At the time I thought it must have been just a slug gun or something, because I was still walking around even though I'd been shot in the head. You see people getting shot in the head on television who die, but I was still walking around and talking—it was pretty weird.

As I was sitting there my head was really aching and I felt pretty woozy. The wound just kept bleeding and, with all the blood, I realized that it wasn't very good, it was pretty serious. The person who lived in the house where I got shot came out and said the offender had pointed the gun at him as well and had then taken off up the street.

Finally an ambulance came and took me to the hospital, and they operated on me that morning and took the bullet out. I've still got it in a container. I only spent about five days in the hospital, but the pain was pretty bad for a while afterward.

On the X-ray, the situation looks worse. The doctors said I was lucky because the bullet only just missed my brain and, because it was so close to my eye, I could have lost my sight too. If the bullet had gone through the eye it could have caused massive damage.

As it was, because of the power of the weapon, or maybe because I have a thick head, the bullet didn't penetrate my skull.

People go, "Geez, you're lucky,'" and I say, "Yeah, well, I was, but still it was pretty unlucky to get shot in the first place." But I suppose I do feel lucky that nothing serious came of it. I'm six-foot-five and people joke that if I were shorter, it would have missed; it would have been just a warning shot over the head of a short guy.

Afterward, leaving the police force didn't really cross my mind. I'd only been a member for less than a year so I was fairly content with the job. I thought of it as the classic scenario—just get back on the horse. I was off work for about a month after the shooting and then did day shift and office work for a while. I got offered counseling but didn't take it up. I felt good about myself. The police force was great; they offered me everything and let me have my space. They didn't push me into anything afterward. They wanted to be quite certain I was all right before I went back into any normal shifts.

A couple of weeks later they caught the guy who shot me. The other guy who was with him had run off onto the beach that night; he'd just crapped himself and gone. Later he told someone about what had happened and word got around, and that's how they got the first guy, luckily enough. He was charged and got nine-and-a-half years. I know where he is now and what he's doing. He's a fairly violent guy.

It was a strange sensation the first time I worked at night again. Walking into the backyard of a house in darkness on a windy night was difficult. It wasn't a hairy situation, but it still felt strange.

> *"I was the guy who got in his way;
> his plan was to kill me."*

Flight attendant Greg Khan foiled a hijack attempt by a man determined to bring down a Qantas flight in 2003. He was awarded a Bravery Medal for his actions.

HOW IT FEELS TO FOIL A MIDAIR HIJACK ATTEMPT

GREG KHAN, *40*

WE ONLY HAD 47 passengers on the flight from Melbourne to Launceston that day. We took off, and I did usual talk about where the bathrooms were and so on. There were four of us on the crew: myself and three girls. I saw this guy get up, coming toward the front of the aircraft. I thought he was going to the bathroom. He got to about row two, only two rows away from me, and I saw that there was something wrong, but I had no idea that I was about to be attacked. He met me nose-to-nose, eyes looking straight at me, and at the same time he'd just gone *whack* into my head. I was pushed back toward the flight deck door. All I saw was something fly past my head that looked like one of those chunky wooden door stops.

I didn't realize till afterward that he had one in each hand and they were weapons he had made himself out of jarrah, a hard red wood from an Australian eucalyptus tree. On each end he'd cut a point like an arrow. When it first went into my head it felt like a closed fist hitting me as hard as it could. I had three women behind me and the flight deck door, and I was thinking, "This is just not going to happen."

He had planned to take the plane down; he had hairspray strapped to one of his legs and a deodorant can in his pocket that he was going to use as flame throwers. I was the guy who got in his way; his plan was to kill me. So I grabbed him by his shirt, put my head down and pushed him back down the aisle. I thought if I could knock him off his feet the attack would stop. The whole time, he was stabbing my head in a frenzy—just continually—he said nothing the whole time, just kept going for me. I'll never forget his eyes: they were wide open and coming straight at me.

We bounced off the seats down the aisle and, at about row eight, a partially blind guy stuck his arm out to try to help. That unbalanced him a little bit and he started to fall, but I wasn't letting him go after getting that far. So we went down, *bang!* with me fully on top of him, but he just kept stabbing me.

Then these two guys who got up to help stepped on the guy's arms. Two other guys dived on my back to try to get him as well. So then they lifted me off and, as I stood up, it was as if someone had tipped a bucket of water over my head—except it was blood. It was running down my face, into my eyes and mouth, just pouring out of me. I slid down the bathroom door and sat on my heels just watching the blood pour from my head onto the carpet.

I didn't even know one of the girls had been hit, but when he came at me with his swings he swiped one of the girls just under the eye and down the face. There was a medic who happened to be on board and he was holding something to my head. I was still in charge at this stage, with people coming up and asking me for things. I asked one of the girls to get the restraining equipment, but she didn't want to open the flight deck door in case there was another attacker who could use the chance to take over the plane. So I got two guys to create a wall across the aisles with their bodies so no one could get past while she was in the flight deck.

I was going to get up and make an announcement, but then the captain came over the speaker to say that we were heading back to Melbourne. The medic got me in a seat, and I said to my brother-in-law, who happened to be on the flight, "Is he restrained?" And he said, "He's not going anywhere." The medic said to me,

"You're going to feel your adrenalin draining out of you now," and it was just amazing, it was like having heat draining out of my head and then I was light-headed, dizzy, dry retching, and all those other things associated with shock. I kept losing consciousness and this guy kept tapping me on the cheek to keep me awake.

We finally hit ground, and the Airport police and Victorian police came aboard to get the guy; it was a relief when he was gone. The captain opened the flight deck door and looked at me with his eyes wide with shock—it really took his breath away, because they had no idea how bad it had been.

The firefighters came on board to get me off and, just as I was getting up, the whole aircraft did this slow clap, like a thank you from them all, and the captain started doing it too. And that brought tears to my eyes. The guy was sent to a mental institution for about thirteen years.

> *"The doctors told my friends I would end up in a nursing home as a vegetable."*

HOW IT FEELS TO SURVIVE A PLANE CRASH

Melissa Holland, *31*

THERE WERE SIX of us, all nurses, on board on the plane. I was the youngest at just twenty-five. We'd been to King Island for a short holiday and were taking off from the island airport to go home to Victoria. As we took off we heard a "beep, beep, beep" and I had no idea what that was. Lucy was the pilot. She swore, then she said loudly, "Put your heads down." I thought it was just a joke.

Janey, the copilot, who was studying to become a qualified pilot, turned around and said to us, "I love you all a lot." And I said, "I love you too." I can't really remember much, but I can remember Lucy saying out loud, "I love you . . " and then the names of her children and husband. Lucy said again, "Put your heads down."

TO SURVIVE A PLANE CRASH

I was looking out the window, but the woman opposite me, Susy, repeated, "Put your head down."

When I realized it was serious, I looked around and the others had their heads down, so I knew that something wasn't right. Reluctantly, I put my head down, too.

Then the only thing I could remember was that something hit my mouth hard and I lost my two front teeth. I don't know if I dreamt this or not, but when the plane crashed I looked up and I remember Janey's arm was twisted behind her back, and then I must have passed out.

We crashed onto the airstrip and a guy from the airport—a fitter and turner—resuscitated me. Lucy, Janey, and Susy all died. Another friend who was with us—she was about sixty-five—went home that night with minimal injuries, but she felt very guilty, thinking that she should have died because she'd already lived most of her life. Another friend who was in the crash, Katy, has had a few back operations since the crash.

I feel a bit guilty too. Susy put her head up to tell me to put my head down, and I think I lost my two front teeth because our faces crashed. She had just gotten engaged.

Janey was in my heart, my closest friend, and sometimes I get really pissed off and I think, "Why did my friends die?" And then I tell myself tomorrow will be a better day, and I hold onto that.

HOW IT FEELS...

After the crash I was taken to the Alfred Hospital by helicopter, with a severe brain injury that left me in a coma for fourteen weeks and in the hospital and rehabilitation for a year and a bit. I died a few times during the first weeks, but I have no memory of it at all.

I only started to be aware of things when I was transferred to Bairnsdale Hospital. When I got there, I was slowly becoming conscious of what was happening, but I had no idea why I was there. I had dreamt that we got home safely. I dreamt so much when I was in the coma. I dreamt I was nursing again and had passed the naturopathic course I had been studying.

I woke up properly when I went to the Caulfield Rehabilitation Centre. A neurological psychologist came and spoke to me. I had an alphabet board to communicate with and I spelled out: "Why am I here?" And she told me I had been in an accident. She said an airplane accident. She told me that three of my friends died. From then on I was determined to get myself moving—to start walking and talking again. The doctors told my friends I would end up in a nursing home as a vegetable, but my friends told my doctors no, that wouldn't happen.

It's been a good battle because I'm still here. I value everything now and truly realize that you only live once, and I appreciate the weather, the water, the beaches, people—everything. I especially value all the people who helped me so much to get through this; it was their support and friendship that enabled me to move forward. But I think I've come as far as I have because of my own determination and courage, too. I want to become 100 percent again. I don't want to be depressed and upset. Because of my head injury I'm supposed to be depressed, but I'm not.

Learning to walk again was hard, really hard. My physiotherapist would tell me what to do, but I would swear at her—I was never nice. My right side is not good at all because of the head injury, and I fractured my back as well. When you have a stroke it affects one side of your body, and it's like I've had a stroke. I also shattered my elbow and left ankle and fractured my ribs. Most days I have pain from my back, and my arm is uncomfortable all the time, but I'll soon be having my last operation, when they'll fuse my elbow more. I don't know how many operations I have had altogether—probably about four on my arm.

I'm happy I'm here. I really believe that I'm here for a reason; I think God wants me to achieve my goals. My girlfriends who died are with me every day; I could cry just talking about it. I really appreciate the people who are in my life now. I appreciate everything around me too, and the good news is I'm allowed to become a nursing aide after I've done a re-entry course. I'm so excited about that. In my head I told myself that I was going to move forward, that I was going to nurse again, that I was going to meet a lovely man and get married. All my life I've had to handle things, and I know the reward is going to be fantastic when I succeed.

> *"I was sad that I hadn't even made it to my twenty-first birthday."*

HOW IT FEELS TO *SURVIVE BACTERIAL MENINGOCOCCAL DISEASE*

PETER, *23*

I'D BEEN OUT with a batch of friends; they came back to my place and we had a few drinks. The next morning I woke up and felt bad, but thought it was just a hangover. I had a really bad headache and stiffness throughout my whole body. I felt quite nauseous and my eyes were really sensitive to light. Just walking down the hallway was a real problem.

I went downstairs and told my parents that I felt pretty ill and needed some painkillers. Nothing really helped, so an hour later I went to the GP. By that stage I thought that it probably wasn't just a hangover, because I've had a few of them, and this was pain I really just couldn't take.

HOW IT FEELS...

After looking at me the doctor said he thought it was just the flu and to go home, have a few more Tylenols and sleep, which was really the worst thing I could have been told. But, you know, you trust your doctor. So that's what I did. I did mention the possibility of a virus or meningococcus, but he said no. Nine hours later it had gotten much, much worse—almost unbearable. Any little movement I made was really painful. I sat there without moving all day, in incredible pain. I had stiffness, and the headache was like a migraine, but much worse than any migraine I've ever had. There was full-body pain as well—pain everywhere.

I started to think about what else it might be, whether the doctor was wrong and what I should do. When my parents came home from work, we decided it was time to go to the hospital. When I started moving and had to switch on the light, then I really knew how intense the pain could be and how sensitive I was to the light. Traveling in the car was quite a bad experience; every little bump we went over made me shout out in agony. I became quite aggressive toward my mom because I was in such pain. Dad told the hospital staff that he thought I might have meningococcus, so I was rushed into emergency.

As I'd been lying down in the dark all day I hadn't checked my body for any rash. The doctor lifted my shirt and I had the beginnings of a faint rash on my stomach, arms, and legs. But it wasn't like you often see on the television, not that dark rash. By the time the rash comes out it can really be too late, so I was pretty lucky that I got to the hospital when I did. They took some blood and gave me antibiotics straightaway. My legs in particular were really painful. I couldn't manage the pain at all, and the doctors told my parents that my legs might need to be amputated.

I was in a delirious state and, hearing that I had the rash, I thought, "Well, this could be it for me." I realized that I could be dying and I was pretty afraid. I was sad that I hadn't even made it to my twenty-first birthday. I had all sorts of awful thoughts about my family and how they might not cope, particularly my parents, or how it would be without my legs if I got through it.

They had to give me a lumbar puncture, to drain the fluid from my brain. I had to be awake for that, crouched into a ball as they put a large needle into the base of my spine. That certainly wasn't pleasant. After that, they gave me a general anesthetic. The next few nights I had very colorful dreams. Dreaming and sleeping offered the only break from the pain.

There were three days of pretty serious monitoring to ensure my condition didn't get worse, then I turned the corner and started improving. The doctors were pretty sure that I would be fine. I still didn't know if I'd recover fully or whether I'd be able to use my legs once

I got better—that was my main concern. I was in the hospital for about three weeks, then I went home and spent the first two weeks in my bedroom, then another two weeks really taking it easy. It was at least a month before I started to go out and see people.

I can't really blame the GP for not picking it up, because meningococcus is so hard to diagnose. I wasn't happy but that's just what happened, and I hope more people and doctors can become aware of the symptoms. When I was in the hospital I remember thinking, "Why is this happening to me?" But now I think, "Wow, I was so lucky," particularly after hearing about a lot of other people who have died, lost limbs, or been disfigured. I'm just extremely lucky, especially as it was originally misdiagnosed, and I did lie around for those nine hours before going to hospital—and that can be too late.

Luckily, I've recovered perfectly now. I can use all my limbs and I've got no scars, though it did take a while to feel comfortable doing everything and I was a bit slow for a few months after that. When I got sick, I was about halfway through a landscape architecture degree and was working in a restaurant. The illness really made me think about life, about where I was and what I wanted to achieve. It was almost like, "I can start again—everything begins now."

I'm now working as a landscape architect and I'm part of the Western Australian meningococcal awareness

TO SURVIVE BACTERIAL MENINGOCOCCAL DISEASE

campaign through the Amanda Young Foundation. I'd never really grasped the fact that I might die, until that happened to me. Now I put a lot more focus on what I want to do; before, I didn't have much goal orientation. The experience gave me more of a drive for life, which is still with me. Life is so precious you really have to make the most of it.

> *"I got washed off I guess about thirty or forty times."*

HOW IT FEELS TO BE LOST AT SEA

Derryl Huf, *62*

WE WENT OUT fishing about 6:15 A.M. on New Year's Eve. We went to a fairly popular place, about twelve miles north of the heads at Port Macquarie. We dropped anchor and the current was violently strong. I baited up my hooks but didn't even put the sinker down. I wound it in and said, "This is hopeless." We couldn't fish in that current, so we went to pull the anchor up, and *bang*! That was it. We were capsized.

I had tried to pull the anchor up by driving the boat forward, but with the current so strong—it was about four-and-a-half knots—I had no hope of making any progress. The sea was becoming rougher; it did change rather rapidly and unexpectedly. Water was coming over the windscreen and canopy, so we had battened down the hatches. Then the anchor rope washed underneath the boat and got tangled in the prop just as a huge wave came, and—end of story: we were capsized. I'd say we had about one-and-a-half to two seconds to do something, if anything.

Our friend Wayne grabbed a life jacket, but my wife and I were just tipped straight in. We had plenty of life jackets and flares in the boat, but there was no time to grab them. I could only hope that we would be able to hold on and keep our strength for as long as necessary in the water.

There was enough rope floating around for us to put it through the eye of the boat. We tied the rope onto that, and my wife held on to it, standing on the bow rail. Wayne stood on top of the hull as though he were water-skiing, holding onto the ropes.

I was sitting between the outboard motor and the boat, in water up to my armpits. I got washed off, I guess, about thirty or forty times. I had a rope lassoed around each wrist, so when I'd get washed off I could clamber back on again. Then a wave would come up about a foot over my head, but so long as I knew when to hold my breath I was coping relatively well.

My wife, standing on the bow rail, was being banged up against the side of the boat, but she held on for the whole twenty-six hours. She was so strong. The water was cold, but you sort of acclimatize to it. We all suffered hypothermia, though.

I think if it wasn't for communication between the three of us during this time we would have lost heart. We were all praying and talking about our children and sort of making plans—you know, that our daughter would have our house, and so on. We thought about our friends and the fact that we'd meet them again one day in heaven. You think of all these things; you think you might die out there.

My wife offered God a challenge at one stage. She said, "OK, friend, if you want us to do more work for you, you've gotta save us." One amazing story was that, just before evening, my wife and Wayne looked up to see a silhouette cloud in the shape of a dove. The dove is the Holy Spirit's sign of peace and was associated with Noah's ark in the Old Testament story. Then the cloud formation changed into the shape of a vessel—a boat. That was fascinating.

During the night, about a dozen or so vessels came towards us, but then turned away towards land, searching for calmer currents, thank you very much. We would yell and scream to no avail; it was very frustrating. The time wasn't right, I guess.

At about 8:30 or 9:00 A.M. a ship virtually came around the corner and Wayne said we should all pray that the ship's captain would steer towards us. Then, instead of going towards the calmer current, he came straight at us and passed within three hundred feet.

Wayne had salvaged a red coat that floated out from under the boat—it was my dear Massey Ferguson coat, which had been caught on something and ripped in half—and Wayne was waving it vividly and relentlessly. Then, when we thought, "Oh well, the ship's gone past," suddenly it slowed and did a right-angle turn. We'd been sighted. And how fascinating—the ship's name was *An Angel*.

It was just in time for me. When I got washed off the boat the last time, I had just about run out of all my reserve power. I thought, "Boy, oh boy, I can't take too much more of this." The ship immediately called the helicopter rescue service; three of them arrived about a quarter of an hour later. What an experience being rescued from the sea! I'd seen it on television, but now I'd been there. I insisted my wife be the first one brought to safety.

The helicopter crew was absolutely amazing; they were so caring. When we got out of the helicopter, do you think we could walk? Not a hope in life. You lose your land legs when you've been out to sea for so long, and it's as though you've had about seven times too much to drink—nothing works. So I was taken to the hospital on a stretcher. Because I was pounded against the motor for all that time I had terrible cuts all over my bottom. My knee doesn't seem to work too well, but I think my sciatic nerve's been fixed somehow. It was an experience that we will never forget and we praise God that we lived through it.

> *"Every drop of air in my body squeezed out of me as seven tons of tractor ran over me."*

HOW IT FEELS TO BE RUN OVER BY A TRACTOR — TWICE

Kim, *Age withheld*

WHEN I WAS twenty-five, my partner and I decided to buy a van and head off around Australia for a couple of years. We took on some work with a farmer for a couple of months. I was working twelve-hour shifts and living on only three to four of hours sleep a night, so really it was an accident waiting to happen. I was helping with the ploughing, towing a couple of ploughs behind two tractors that were hooked up together, one behind the other. It was a bit of a challenge to drive them and it was really dangerous; you wouldn't get away with that sort of thing today.

The farmer had given me a few lessons driving them, but on my first go I stalled it, and the second time I jackknifed it! I told him at least I could only get better, but he didn't see the humor in that. I eventually got the hang of it, did a quick lap around the paddock, and the farmer was like, "You'll be all right, friend."

This day I was driving the two tractors as usual, with two ploughs hooked up behind them. I finished one paddock and jumped off the front tractor to make sure the ploughs were out of the ground before I continued. I had to go around to the back of the tractors and manually lever the hydraulic pump to lift the plough out of the ground. I was standing there operating the pump when I heard the tone of the engines change a bit and realized that the clutch of the rear tractor was engaging. I thought "Shit, that tractor is in gear."

I ran around to the front tractor, to try to reach in and grab hold of the clutch lever to disengage it, but I was too late. I was standing in front of the tractor's five-foot-high and two-foot-wide dual wheels when the whole show just started moving. Once the clutch gripped in the rear tractor, which was fully revving, it just took over.

I tried to take a small step sideways, but then I was flat on my back and the thing I remember most is just the incredible rush of air as every drop of air in my body squeezed out of me as seven tons of tractor ran over me. It was like the air rushing through a blowhole. That's something I will never forget, I can promise you.

TO BE RUN OVER BY A TRACTOR — TWICE

The wheels were about as wide as my body and when I came out from under them I was somehow flipped onto my side. There was only about ten feet between the wheels of the first tractor and the second, so there was no time to think "this is going to hurt" before the second tractor was on me.

All I heard was this noise like a rifle shot as my collarbone snapped, but there was still no pain; no thoughts and I still had the plough to go through. The front wheel of the plough, called the furrow wheel, drove over my head and pushed into the dirt, and it wasn't good, but it was nothing compared to being run over by two tractors.

By this stage I was feeling a bit dazed and I heard the thud and the bang as the plough discs ran over me. I couldn't breathe, I had dirt in my face and in my mouth and my head was pushed back into the ground, but I was still the calmest guy there, I wasn't panicking. Anyway, I didn't have too much of a chance to lie around feeling sorry for myself. The next thing I knew the tractors were coming back to have another go at me.

They came around in a circle and I thought I was going to be run over again. I staggered up, I could hardly see because there was so much dirt on my face, I tried to move away from the tractors as best I could. I went on for a bit, zigzagging back and forth while the tractors went round in circles down the paddock and eventually jackknifed and stalled.

I was more focused on the fact that I still could hardly breathe. My partner happened to be looking out of the window of the van, wondering why the tractors were going around in circles and why I was staggering around. When I fell over, she came running out. I could get a few words out by then and told her what had happened. She called up the farmer and he came screaming up in great panic.

I'm lying in the dirt thinking he's going run me over too, just for good measure. His bloody dogs leapt out of the ute (utility vehicle) and were jumping all over me and licking me, which I could have done without. He was all for throwing me in the back of the ute and driving me to hospital, but "Not on your life," I said. I thought I'd had enough excitement for one day, and could just imagine getting bounced out of the bloody thing on the way there, so I told them I'd just wait for an ambulance.

It took about an hour to get there and in that time I did wonder if I might die. When no one was looking, I wiped my hand over my mouth to see if there was any blood, because I thought I might have internal bleeding, but there was nothing. By the time the ambulance got there I was a bit sick of the whole thing, really. When they picked me up there was a dent where my body had been pushed into the dirt.

That night I was sitting up having a bite to eat in the hospital with everyone wondering how on earth I'd

TO BE RUN OVER BY A TRACTOR — TWICE

lived through it. The next day I wasn't so chirpy, though. I couldn't move, I just lay there and hurt. There was a huge bruise on my chest with a big V pattern from the tread of the tires. I spent two weeks in the hospital, but the only serious injuries I had were a broken collarbone and a broken bone in my foot. My knees were never the same again, however, and now years later I need a double knee replacement.

I did have bad dreams of being squashed afterwards, but I still went back to driving tractors, though sometimes I would have an involuntary shudder if I looked down and saw mud squelching out from under the tires.

> *"I had a clear image of myself lying under the water, lying there dead with my eyes open."*

HOW IT FEELS TO DROWN

Sue Merrotsy, *41*

I WAS SURFING as normal one day. I got off the board in chest-high water and went under. The other guys surfing with me thought I'd gone under to look at fish. When I didn't resurface after about four minutes, one of them paddled over to me, by which time I'd been under a few more minutes. He pulled me out and dragged me to shore.

I was gray, not breathing, and my lungs were filled with sand, and there was no air in them. I was dead for about ten minutes. The guys did CPR, then a couple of doctors came to help. The paramedics arrived and used a defibrillator to shock my heart. My throat had closed up, so they had to put a steel thing in it, and an air tube to pump air in. They got my heart going after the third shock, then the helicopter arrived and they flew me to the hospital.

When they started my heart on the beach, apparently I said two things: "It wasn't my time to go yet," and "The videos are due back." I don't remember any of it; friends have since told me what happened. I lost two days. I don't remember going for a surf, I don't remember anything.

I was on life support for three days because my body had shut down. The doctors told my children that I had an 80 percent chance of dying. My sister-in-law even sent for the priest to give me the last rites. They said my brain was twice the normal size because all the water got in and swelled it; the doctors said I had an excellent chance of brain damage because of the lack of oxygen.

I woke up three days later in the wards. When I first woke up it was very hazy, but within a few days I was back to normal and had no lasting damage. The doctors have said my survival was a miracle and extremely rare—to be gone for ten minutes, come back, and suffer no visible effects.

No one knows why I went under the water. People suspected I might have hit my head, but I had no abrasions. They did tests a month after the accident and said I had the heart of a twenty-year-old. About four weeks after the accident I went for my first swim, and I don't know if this is something I created but, when I went under, I had a clear image of myself lying under the water, lying there dead with my eyes open, very aware and just feeling so happy. I can't explain it. I've never felt such happiness and peace. When I came out of the water again I started crying; I was really angry and upset. It was like, "Oh that was so nice."

I can't explain how it happened. I do know that before the accident I was in a really abusive relationship. I'm talking really abusive; physically and mentally. I'd gotten out of that finally, after about two-and-a-half years. After the accident my counselor suggested that maybe, because the guy still kept harassing me, a part of me just lost faith, thinking I would never get away from him, and so that part of me just thought, "Well then, fine—I'm going."

I have to give some credence to that possibility, because I remember that I felt powerless and I didn't know what to do any more and I didn't know who to speak to. I didn't know how to stop this person. All the stress of going through that may have created a high amount of adrenaline in my system and that could have been the reason I just slipped underwater.

It wasn't intentional, but when you have been through years and years of abuse you feel very powerless. I'm educated and intelligent, but when it comes to things like that it doesn't matter who you are, you can get caught up in a cycle you don't know how to get out of. There was also the chance that it was a heart virus which can just come and go, boom, like that.

I really think it was a big shake-up for me, not at the exact time of the accident, but in the six months afterwards—that's when I decided I want the best of everything, that's when I met my new partner, that's when I got accepted to study law, that's when I started a stained glass window business, and I love doing that.

After the accident the local surf club raised five thousand dollars for my family. It made me feel really loved and part of a community. It was lovely—there was money left in envelopes at our door, the lawn was mowed and I didn't know who had done it—those sorts of things. The accident showed me how wonderful strangers can be, how wonderful friends and community can be, and, you know, if they could all support me, then what the hell was I doing not supporting myself?

It was three months before I went surfing again. A month after the accident they put a defibrillator in the muscle above my left breast, so if my heart stops or goes into arrhythmia, this will restart it. It's a seventy-thousand-dollar precaution. It looks a bit like a third breast because it sticks out, but the deal I've made with the doctors is that if it stays there for five years and doesn't go off, then they'll take it out and I won't have to have it replaced. So far my heart has behaved perfectly every day.

HOW IT FEELS...

These days I spend about four to five hours a day surfing. I have my dogs, I go for walks, I study, I've got my stained glass business, I read and I listen to music, but surfing is my real passion.

> *"I was too sick to worry about whether I was dying."*

HOW IT FEELS TO
CONTRACT DENGUE FEVER

Emily, *38*

IN 1989, I went for a three-week holiday to Thailand with a girlfriend. We flew into Bangkok and had to trek to get to our destination of Ko Somui, which was a really out-of-the-way place in those days. We didn't have any accommodations booked; we just went on an adventure to see what would happen. We had had all the right immunizations and so on before we left.

The first night we were there, we stayed at the first bungalow that we could find. The next day we set out to find a better place. We got lost on this island, where there was no transport. We met a Thai guy who invited us to stay at his place; he also lined us up with a French Naval Seal who had a Japanese sailing junk—a beautiful,

beautiful boat. This man organized for us to go to a local island and travel around with him, and that's where I got dengue fever on an island called Ko Phangan. It was a very big druggie island, but we didn't know that.

When we came back from there I was covered in welts from mosquito bites; it was the only day I got bitten. Seven days later we flew out and I was starting to feel a little unwell. I think if I'd got sick in Thailand I would have been as dead as a doornail pretty quickly. We got home at 7:00 A.M. and that night I was supposed to do a nursing shift, but I had started violently shaking. I thought it was because I was cold after coming from such a hot climate. I went to my GP and she said I had the flu. I asked for a blood test, but she refused and sent me home.

That night my mom went out on her first date with the guy who is now her husband. I said, "Go out, I'll be fine." Two hours later I rang my dad, because all my joints had started freezing up. It was like a rheumatic freeze in the joints—I couldn't move.

They got me into hospital, but the doctors had no idea what was wrong, and it was just awful. My eyes bulged out of my head and I couldn't stop shaking.

For six days they didn't know what I had. I was so photophobic they had to nurse me in a black room. I had an endless string of doctors and specialists coming in to see me, and it was terrible, because they also had all the medical students, everybody, in there. I had so many people around my bed poking and prodding, and so many blood tests.

I couldn't move because the headache was so severe. My cell count was pretty much nothing at that stage and every part of my body was frozen. My platelet count dropped, and then I started to hemorrhage from every organ, every part of my body. If water touched my skin it was like boiling hot acid attacking me.

I lost almost all my hair—it just fell out until I had about 20 percent of it left. I lost a huge amount of weight and looked just disgusting. I was completely covered in a dreadful red rash and had the worst pain in my eyes you could ever imagine. It was horrendously, horribly painful, but I was too sick to worry about whether I was dying or not.

I was also concealing another problem from the hospital staff, because they had all these young, really gorgeous doctors. As I was coming out of the fever and was able to open my eyes and look around, I saw all these stunning men coming into my room. I couldn't tell them that years ago I'd had a horse-riding accident, and now, because of the fever, a major hematoma was forming between my legs from that previous injury. I had like three massive sausages swelling in my girly bits. I sat on that for weeks in the hospital, just swelling and swelling. I was too humiliated to tell them, even though I was a nurse and I knew what was happening.

HOW IT FEELS...

When I was finally discharged from the hospital I got into the car with Mom and said, "You've got to take me to the GP." The GP just looked at me and said, "Holy shit, you need to be in a hospital right now." But even he grabbed all his associates and said, "You've got to come in here and look at this, so you'll know what it is if you ever come across it."

I was readmitted to a different hospital that night and was operated on by a gynecologist and had everything reconstructed down there, which has left me permanently impaired. They had to drain it all and rebuild it. I've been left with changed anatomy.

After I left the hospital, the infectious diseases doctors called me and said what I'd had was hemorrhagic dengue fever. It was six months before I felt normal again. I missed a semester of university because of it. They told me the fever would never come back, but it did come back. I spent a few years with it coming and going, when I'd start shaking again and feeling sick. But it seems to be over now.

> *"I saw this blur, a flash of teeth and water, as it ... took me down for a death roll."*

HOW IT FEELS TO BE ATTACKED BY A CROCODILE

VAL PLUMWOOD, *Age withheld*

I WAS IN a canoe on a tributary of the East Alligator River in Kakadu, looking for Aboriginal rock paintings. I had been out the previous day and it had been idyllic. This day, though, there was drizzle which progressed into torrential rain.

In the afternoon I had a strong feeling of being watched, and suddenly the canoe seemed very flimsy. I had a sense of danger or vulnerability and decided I wanted to get out of there. I started paddling down the river and hadn't gotten very far when I saw what looked like a stick ahead of me. As I was swept towards it, I saw eyes in it and realized it was a crocodile. From what I could see, I thought it was probably a small one, so at first I wasn't alarmed. I'd seen small ones on previous days. It turned out that I was wrong and it was probably about eight to ten feet long.

I was almost past it when there was this great blow on the side of the canoe. I started paddling furiously but it followed, bashing on the canoe. I started looking for a place to get out, but couldn't see one. I felt sheer terror. This wasn't something I'd never thought about—but I'd been led to believe that canoes were safe.

I saw a tree growing from the water near the bank and thought maybe I could leap into it. I stood up and got ready to jump, and as I did so, the crocodile came up close. I looked straight into its eyes and it looked straight into mine. It had beautiful golden-flecked eyes. I remember those very strongly. I did the thing you're advised to do, waved my arms and shouted and tried to look fearsome.

Then I jumped, and it got me in mid-jump. I saw this blur, a flash of teeth and water, as it grabbed me between the legs and took me down for a death roll. I thought, "I'm not food, I'm a human being, I don't believe this." It seemed as though I'd entered some sort of parallel universe where I didn't matter—I didn't count for anything. Suddenly I was in a universe where I had no more meaning than a piece of meat. That's a hard thing to accept.

There was incredible, searing pain, but I was mostly concerned about the roll, which seemed to last forever;

it pushes water into your lungs and you have a strong sense of suffocating. It felt like my arms and legs were coming off—a great sort of whirling sensation.

My head came up above the water and I coughed the water out of my lungs and started to howl with pain. Then it pushed me into the second death roll. We came up again, and this time right next to me was a big, solid branch, so I grabbed onto that. I hung on grimly, thinking I'd sooner let it tear me apart than go through another death roll.

Then suddenly I felt the pressure relax and I realized it had let go. I tried again to jump into the tree. This time it grabbed me around the leg. Its mouth was right around my leg and up around my hip. It took me down for a third death roll. Again I thought I was going to die.

I thought it was going to take a long time over it, which seemed worse than having it kill me straight out. But a minute later it let go of me again. I tried to throw myself at the mud bank. I got almost to the top and then slid back down again. I did it again but didn't get as far. As I slid down the mud I managed to grab a tuft of grass and hang there. I found, by doing that, I could push my fingers into the mud and climb the bank.

I got up to the top and stood up and couldn't believe it; I was still alive. It was an incredible rush of elation. But because I was still in danger, I floundered on. My leg wasn't working very well. I went on for some time before

HOW IT FEELS . . .

I stopped to look at my wounds. They were terrible, much worse than I had anticipated. The leg wound was particularly bad, with bits of tendons and stuff hanging out. I had shock right through my body and was feeling pretty sick. I tried lying down but felt worse, so continued to walk back in the direction of the ranger's station. I felt just a glimmer of hope that I might survive.

The rain was still torrential, and it took me a couple of hours to reach the lagoon I would have to cross to get back to the ranger's station. At this stage I started to black out, and had to crawl. But then, all of a sudden, the rain stopped and the whole area became still, abnormally still, and so the ranger was able to hear me shouting for help. It was then a thirteen-hour trip to Darwin Hospital, and I eventually recovered after almost a month in intensive care.

It was really a life-changing event for me. It was quite a while before I took in the full extent of how it changed my way of looking at the world. I have an incredible sense of feeling that life is not something to waste. I have a real glow and sense of gratitude about being alive, which has never left me.

The experience also changed my outlook. I had a sense that the experience was a dream, but I now think it's ordinary life that is the dream. We don't understand that we are ecological beings who are part of the food chain. We're still fighting that.

TO BE ATTACKED BY A CROCODILE

The experience also had a big impact on the direction of my work. Afterward, I started writing about how we see ourselves as outside nature. I'm writing about the power of nature and our illusions, how we need to see ourselves as being inside of nature.

Your final moments, those life-or-death experiences, have an incredible intensity. That's why they have the capacity to change life. You see things at that point which you wouldn't normally see; it strips away a lot of your illusions about your life.

> *"I saw two hairs curled like springs and stuck on my penis."*

HOW IT FEELS TO BE ABDUCTED BY ALIENS

PETER KHOURY, *Age withheld*

MY FIRST CONSCIOUS recognition of being abducted was in 1988. I had just moved back in with my parents, and one night I went into my bedroom, switched on the TV, and went to lie back on the bed, but my head hadn't even reached the pillow when I felt something grab me around my ankles. I instantly felt pins and needles crawling up through my legs, up through my thighs, and right through my body to the top of my head. When it reached the top of my head, it just felt like static electricity or like thousands of ants crawling through my scalp, and instantly I became paralyzed.

I couldn't move. I tried to call out but I couldn't

make a sound. I was thinking, "Call out Sam's name!"—that's my brother. But I couldn't. I started to think, "Oh my God, I'm twenty-three years old and I'm paralyzed. I'm going to be in a wheelchair for the rest of my life." I wasn't thinking aliens or UFOs—nothing like that. I thought it was some sort of disease. I thought it was payback for a fairly wild lifestyle. That's when I started to get really frightened.

Then, all of a sudden, I look at the end of the bed and there's a creature there, a being. Horrific-looking, at that. There were either two or three of them between the wall and my bed. I realized one of them had grabbed me around the ankles. I thought, "This is evil, this is the devil. I'm dead; they've taken my soul." On the left-hand side of the bed were two beings totally different from the others. The first lot were about three to four feet in height. Imagine a gorilla's head that is very dark, with wrinkly, shiny, leathery skin. They had very big, round heads and were wearing robes with hoods, which made their heads look even bigger.

The other two were seven or eight feet tall. One was a female, but they looked almost identical. They were wearing some kind of jumpsuits, with what looked like doctors' white coats. Both of them had surgical masks around their necks. They had oval-shaped heads that came to a point at the chin; their eyes were very, very big, cowlike, and almond-shaped. Their noses were very small and long. I couldn't see any teeth or tongues—their mouths were just little slits.

The one closest was leaning over me. As soon as I looked in his eyes it was like I fell into a void and it just overwhelmed me; it was like he knew everything about me. There was a connection, as if he were speaking to me through his eyes—it was telepathic. He was "saying" to me, "Calm down, we're not going to harm you, it'll be like the last time." That made me think, "What last time? What's going on here?" He had in his hand what looked like a syringe, but more like a crystal wand—a little tube with a tiny light at the point of it. He pointed it to the top left side of my head and pushed it in. I didn't feel any pain, but I blacked out.

I woke up and sprang out of bed, and the paralysis was gone. I thought I'd been asleep for only about fifteen minutes, but when I looked at the time it was 2:30 A.M.

The only memory I had was of being in a round room where the walls emitted light. It was like being inside a pearl. I remembered lying on a table and someone standing above my head and speaking like a bird chirping. It was one being, but it sounded like thousands of birds. I was wondering, "How can I understand what you're saying when you're talking at this speed?" and he said to me, "You will remember," but I can't.

After a shower the next morning, I felt burning on my

right shin; there was this red, raw mark as though a cigarette had been left to burn on it. Six hours later I went to pick up my wife and I told her that something had happened. I went to show her the burn on my leg but it was healed. There was still a mark there, but it had healed after only six hours.

It's a deep, round scar about a quarter of an inch in diameter and it has three marks on it, like a triangle, in the middle. It's called a scoop mark and is common among alien abductees. I've got about four or five now, but that was the first one I ever got. I also had a scar on my head like a doughnut; it was raised rather than indented.

I think what the beings are doing is treating humans the same way we treat animals facing extinction—catching them and implanting them with a tracking device. I think they need to track us. I have no doubt that a lot of people are being tracked. I have no doubt that I'm being tracked. My son also has an implant in his hand. You can see that there's something under there, but we don't want to say much or do much about it. He's OK, my kids are OK with it, they're not scared or anything. I've never been fazed by what's happened to me. It's had a big effect on my life, but it's been a really positive one.

The next abduction was in 1992. You'd think it couldn't get stranger than the first time, but it did. This day I drove my wife to the station and, on the way, I must have vomited about eight to ten times. I had to pull over each time and was violently ill. I said to her, "Where is this coming from?" On the way home it was even worse. I came home and went straight to bed. All of a sudden it felt as if a cat jumped onto the bed. We didn't have pets at the time. I opened my eyes and, straddling me— sitting on top of me—was a naked female. Another female was sitting on the end of the bed.

The one sitting on me was a humanoid in that she had arms, legs, and hair. She had human features—eyes and nose, mouth, everything, but the eyes were more than twice as big as human eyes, and her nose and face were longer, more stretched. I was still fully dressed but they were both naked. One of them was completely blonde and the other one was Asian, very petite, with a bit of a belly. She was sitting on the corner of the bed with her legs folded under her.

I was stunned. I thought, "What the hell? How did they get here, what are they doing in my room?" I thought someone had broken into the house and tied me up. It took me half a minute to realize that it wasn't a dream. This was really happening. Next I sat bolt upright and, as I did that . . . this is very hard to explain, but I'll do my best . . . it felt like my soul sat up quicker than I did, and as my body tried to catch up with my soul, I could see myself. There was a kind of transparent image of me. It was really weird.

TO BE ABDUCTED BY ALIENS

Then the female cupped her hands around the back of my head and brought me to her breast and buried my face in her breast. I'm claustrophobic at the best of times, so this was awful. I pushed away with both my hands against her body but she pulled me toward her again. The third time she buried my face in her breast so hard—she had super strength—that I couldn't breathe.

I don't know why I did what I did next. I've never done it to any other woman and I don't think any woman would appreciate it, but I took a little bite of her flesh. I think I bit the nipple and a little bit of skin came off and went in my mouth and got stuck in my throat. She pushed me back, looked at me, looked at the other one, and I could tell they were communicating and she was saying to her, "This is wrong, it didn't go as planned"— that sort of thing. I got the impression that they had done this with me before.

It was definitely a sexual thing. I mean, she was naked and straddling me.

Because I'd swallowed this thing I got the worst coughing fit you can imagine. It was like a chemical reaction in my throat. It was burning the hell out of my throat. I couldn't believe how bad it felt. I'm coughing and coughing, and I look up and they're gone— vanished.

I left my room, went into the kitchen and ate some bread, and tried to wash this thing in my throat down with water. It was stuck in my throat for three days.

Afterward I wished I had vomited it out or coughed it out. It might have been good DNA. People asked me later if I checked for it in my bowel movements, but that never crossed my mind.

Then I had the urge to pee, so I went to the bathroom and, as I was about to go, it felt as if barbed wire or razors were ripping at my penis and . . . it's a bit private . . I pushed my foreskin back and I saw two hairs curled like springs and stuck on my penis. It was really, really painful to remove them because they had dug in. It was like they were hooked in. One was six inches, the other about three inches. We don't actually know if they were from the head or the pubic area of the beings.

I think whoever was in my room were hybrids or clones. I don't think they were fully human because when I bit on that nipple she didn't scream and she didn't bleed. Her skin was very, very pale, as though she hadn't seen the sun. Her hair was sort of blown back, like Farrah Fawcett's when she was in *Charlie's Angels*.

At one stage one of the beings touched her stomach and I think she was pregnant. I believe they're trying to replenish their race because they can't actually reproduce, and whether it's because they don't have males, or their males just don't have any sperm, I don't know. I think they harvest sperm from us, and are building up their population or trying to produce a hybrid, someone who looks more human. I think they were using me for reproduction and that they had done the same thing to me before.

I kept the hairs and had them scientifically tested. The scientists tried to photograph them, but because they were transparent, they had to spend about ten thousand dollars to get a photo done with special microscopes. You can only just see them with the naked eye. The scientists found that the hairs have a mosaic structure that our hair doesn't. They concluded definitely that it's not human hair—only two or three percent of the creatures on this planet have a similar DNA structure. It's been listed as the world's first biological evidence of an alien abduction case. So it makes the case pretty big.

The aliens will come back to me, without a doubt. All I wish for is the opportunity to ask some questions. I want to be a willing participant and I'd at least like a say in the matter. I think their world is more dimensional than just simply being about space travel. I think it's right there, next to us, but it's in another dimension and they know how to manipulate it.

> *"I can channel all animals. I've done it with a sheep which thought it was a horse."*

HOW IT FEELS TO BE
AN ANIMAL PSYCHIC

AMANDA DE WARREN, *Age withheld*

ABOUT THREE YEARS ago I went to a lady's house to do a channeling and her German shepherd kept trying to get my attention. I was looking at him and thinking "that dog is trying to tell me something." Then all of a sudden the dog went, "I was abandoned by my first owner who left me in a cage. I'm sorry that I chewed up the white blanket, I'd like my walks to be longer, and I don't like the chicken stuff my new owners give me."

When I told the lady who owned him her jaw nearly dropped to the ground, because it was all exactly right. So then I started channeling animals and found I have a direct link to them. It's almost the same as human channeling in that I see telepathic pictures that they show me and experience senses and feelings they give me. The animal doesn't have to be alive; it can be passed on, and it doesn't have to physically be with me or the owner when I do the channeling. I can be sitting talking to the owner while the dog is running around the backyard. Normally people send me a photo and we can do it over the phone.

I can channel all animals. I've done it with a sheep which thought it was a horse. I've done birds, and I had a fish come through to a lady the other day and I said to her, "You killed your fish didn't you; you put hot water into its tank." She said, "Yes, I did." We had a laugh about that.

I haven't tried to channel an insect. I guess I could, but it would be very interesting. I've done spiders, though. This boy had a pet spider, and his house had caught on fire and the family couldn't get it out, so it perished in the fire. The spider came through and had a chat.

I'm able to tune into particular animals when I need to. It's not as though I go to the zoo and hear all these voices. But I did help one zoo, and I'm not allowed to say which one, but they had a problem with an elephant and they asked me to channel into it and find out what was going on. The elephant told me that it didn't want to eat in front of people, so the zoo changed the feeding time and the problem was solved.

HOW IT FEELS . . .

I'd like to speak with whales about why they beach themselves. I think it could be such a bonus for science to use my skills. What I do helps people to understand their pets a lot better. Instead of putting them through expensive and painful tests, I can immediately see the problem. They might tell me that they have a sore tummy or they don't like their food or whatever.

Animals mostly want to talk about what they're fed, what they like and don't like, or any ailments or pain they may have. I can help with behavioral issues as well, because I connect with the animals and understand what they're trying to say; sometimes they don't understand that they're not humans and I have to set them straight.

People tell me that after I've communicated with their pet it helps to set the animal right—I can tell them that it's not OK to pee through the house or chew electrical wires or whatever. A lot of animals also want to reassure their owners if they were responsible for their passing over—if they were put down. It gives people a lot of comfort to know this, and that's a big reason why I do this work. I believe that animals have souls and I know that our pets wait for us in heaven.

A few months ago a lady called me because she was having trouble with her dog and I was able to see that its bladder had been damaged in a recent surgery and had been leaking ever since. A really strange thing happened then; I felt my soul going into the dog's body and I pinched the bladder where it had been cut and later I heard from the owner that the bladder had not leaked since then.

I've also recently spoken to a dog who witnessed the murder of his young owners—it's quite an infamous murder case at the moment—and the dog has shown me some of what happened, but the police seem ready to charge the wrong person. I've told this information to the people involved, but I don't know if it has been passed onto the police.

I now have up to sixty animal-channeling clients a week; it's becoming more popular than human channeling. It's quite important because I don't think there is anyone else who can do what I do. I don't eat animals since discovering this ability. I used to, but one day the lamb I was eating came through to me and ever since then I haven't been able to eat meat.

> *"I proved I was not vegetative, but locked in."*

HOW IT FEELS TO BE "LOCKED-IN" (TO BE IN COMPLETE PARALYSIS)

Airlie Kirkham, *38*

I WAS A Japanese language teacher in Adelaide. I had a very satisfying and happy life. I loved teaching, enjoyed music and friendships. I had a wonderful boyfriend and friends from university where I had completed my arts and music degrees, and a graduate diploma in teaching.

When I was twenty-five, I was driving along a country road to a kitchen tea party. I somehow found myself on an unknown dirt road. Going round a bend, I skidded into the path of another car (I was not speeding, but my wheel may have caught the gravel on the edge). I was in a coma for several weeks with a very severe brain injury and several broken bones, and I had my spleen removed. I was not expected to live. After the coma, I

TO BE "LOCKED-IN" (TO BE IN COMPLETE PARALYSIS)

was very sleepy, but I remember my parents sitting by my bedside every day and night. I wasn't in pain.

Eventually I went to the Julia Farr Centre to begin rehabilitation. I was there for over six years. I recall big wards, many people who rushed around looking too busy to be friendly. The rooms were full of people like me, accident victims who were unable to do more than pass the time.

My time seemed to drag. There was little to do except watch TV, and I couldn't see that properly. They put us to bed so early. How could we complain? Few of us could talk. At times people would come and talk over me, not to me. Many thought I was vegetative, because I could not communicate or respond, or even move a muscle. I could move my right leg a bit but that was all. I felt locked-in because I couldn't speak or even write.

The "locked-in" state is a diagnosis applied to people who demonstrate alertness and wakefulness but who cannot respond. I could see everything and everyone. I knew who they were. I could hear people talking about me and saying things I didn't like.

Since I could hear and understand, I would have liked people to talk to me properly.

I felt very frustrated, at times angry, because often no one would come near me or talk to me—only Mom and Dad, my visitors, or therapists. I was frustrated because I couldn't make decisions, or tell people what I wanted.

I wasn't in control of myself. I learned to wriggle my finger, hoping someone would notice and talk to me. I wanted to tell everyone, "I'm still here. Inside me is Airlie. I have so much to say."

The seasons came and went. I made some progress but I was unable to assert myself or respond. I was bound by ritual, routine, habit, culture. I set my sights on being able to communicate. Couldn't I do that? I had been an articulate person once. I could be that again.

After about five years my right arm recovered some movement. Then a miracle occurred. At Mom's request, the occupational therapist was asked to try and help me write again. She made me a special penholder and glove. The first word I wrote was "Airlie." Then I answered a question from the therapist who asked if I wanted to learn to write again. I wrote, "Yes. I need your help."

At first I wrote like a kindergarten child. It took me over six months to learn writing again and I have not stopped since. The pen and its holder are my lifeline to the world. I proved I was not vegetative but locked in. Now I was free at last. I felt I had the whole world opening up in front of me.

I wrote down all my questions. I had so much catching up to do. I felt wonderful now that I could talk to Mom and tell her all my problems and feelings. It was like a door opening, and all my thoughts came pouring out. I felt so excited and happy now that I could tell

TO BE "LOCKED-IN" (TO BE IN COMPLETE PARALYSIS)

everyone that my brain could still function, that I wasn't stupid or unable to understand. Even today, people often say to Mom, "Can she understand what I'm saying?"

I felt so much love in my heart for God, who had answered my prayers for survival and had helped me communicate again, and for my family who had looked after me so well. I turned to writing poetry. Poetry has always inspired me.

To other families with relatives in a similar situation, I would say, never give up hope; be strong and determined. You need to communicate to your loved ones because even people in a coma can hear and think. They are sensitive to their environment and need to be aware of their family's strong hope and faith.

In 2002, I enrolled in a refresher subject for a Bachelor of Music and, with the support of my carer and mother, achieved a distinction. In 2003, I enrolled part-time for Bachelor of Music, Honors Musicology, and achieved good results. I still need total support for daily activities. I still cannot walk or talk, but I can think and communicate, using my pen and a delta talker machine which speaks for me. University is the best thing in my life at present. What the future holds I don't know. The path of life that I tread now is not the path I would have chosen for myself. But I will walk it, with the strength of God whose steadfast love has never failed me.

> *"It's not so much the first one who dies; it's the last one left alive."*

HOW IT FEELS TO BE
AN IDENTICAL TRIPLET

MARTIN MCKENNA, *42*

WE WERE BORN in Limerick, Ireland, in 1962, not very long after the Second World War. The people of Limerick were shocked when suddenly, this six-foot blonde bombshell appeared in their town and had these blonde triplet boys, each with one green eye and one blue eye. Being Irish, the townsfolk put two and five together and came up with the idea that we were part of the master race, Hitler's plan for genetic purity.

People used to ostracize us. They had a deep-seated hatred of Germany. They would paint swastikas on our wall and refer to us as "the Nazis." It was very sad, and it was very hard for my mother, who had been a refugee. Irish people can be afraid of anything different, and yet three is the luckiest number in Celtic mythology. We could have been seen as an omen of good fortune, but they saw it the other way.

Genetically, identical triplets are very rare. There were three of us in one sack inside my mother, each fighting for our share of everything. I was the smallest, weighing two pounds, one ounce. My mother didn't even know she was having triplets because they didn't have all that fancy equipment in those days. So two boys were born and they put her back on the ward. Then she started complaining about pains in the stomach so they pulled another one out and that was me.

We had no immune system so we weren't handled by humans; they put us into these incubators. We didn't get touched until we were about five months old, so when we got home we wouldn't let our mother hold us—we would scream and struggle. We would cuddle together, the three of us, or if the dog came in we would cuddle with him. My mother always said that when we came home from hospital she was going to give us back; she thought there was something wrong with us.

When we were born it was a big hurrah. All the newspapers came around. A baby food company sponsored us, but we were allergic to the food. I don't know the medical statistics of genetic triplets, but I do know that when I was a kid, there weren't any other triplets around. We used to walk around and we used to say to each other, "There's nobody else like us, is there? We might as well come from Mars, hey?" Our neighbors on the right side had twins, and on the left side they had two sets of twins. There was a fairy mound, a very powerful one too, behind the house. Maybe the fairies had something to do with all the multiples.

When we got older we would never go out in public because of the people pointing at us all the time. We spent a lot of time in the house and no one would take us out because we'd freak out. Even our first day of school was bad. We barricaded ourselves in the school and we wouldn't let anyone near us. The teacher said, "There's something wrong with these kids."

When we got baptized at the age of seven, well, we didn't know what a priest was! We were in the church and there were all these horrible pictures in there. We thought, "Jesus Christ, what is this about?" And there was this big basin full of water and a guy came out in a cloak, so we were very wary. Then he picked up John and put his head near the water. Well, we thought he was going to drown him, so we went up and knocked the basin over, knocked the priest over, and ran out.

We didn't talk for a few years because we had our own private language going; we didn't need to let anyone else into our little club. I can still talk it today, and even now, when we get together, we use it. We laugh and we're just like kids again. When we were growing up, if any one of us had to get spanked, we all got spanked; if anyone was getting a reward, we all got a reward. My mother called us "johnmartinandrew."

When I was about nine or ten I was a really bad little guy and I was pretty anti-Christian after that episode in the church. My school tried to break me. But I always knew when I was in trouble I could go to our dogs, two German shepherds that my mother loved. One day I got in some serious trouble at school so I ran home to the dogs, but the principal followed and I set the dogs on him. The next day the dogs had to be destroyed for being vicious, and I knew that was it for me with my mother; she wouldn't look at me after that.

I ran away and lived in a little cowshed. There were lots of kids living on the streets. It was a pretty dangerous place, so I ganged up with a group of old dogs, because I was pretty good with dogs. I didn't trust people because they all let you down, but I had my pack of dogs. I used to sit down and watch what they did, and I could understand their communication. There were no fancy words. The rules were dead simple. I thought, "That'll suit me, to live like that," and I stayed for three or four years. My brothers would come down and bring me food. Me and my brothers would fight with the other street kids for a place on the back of the rubbish truck as it headed for the dump; we'd get our fruit and stuff from there.

Being a triplet has been a blessing, though. It makes you think of someone outside of yourself. I've been thinking about and communicating with another two people since before I was born. At fifteen, I broke loose from the circle. We were apart for about five years before I realized that I was just me, I had to stop saying "us" or "we."

When we were in Ireland everyone used to point their fingers at us and say we were freaks, but after we left and met up together in Australia or the US, people would marvel at us. They didn't call us freaks. We're tall—over six foot three and around 195 pounds—so we make a striking picture.

I came to Australia with fifty bucks, and I've had a great life. For five years I've been doing radio shows for the ABC as "The Dogman," a dog behaviorist. I've published a book, even though I can't read or write, and I'm about to start a documentary. I have a beautiful wife and beautiful children, and I don't think I would have achieved all this anywhere else.

John now lives in South Africa and Andy went back to Ireland. Splitting apart was hard. We used to say, "What can we do? How can we break the magic circle? Are we going to be old men, three of us living in a house together?"

But even now, if one of my brothers is seriously ill, I know it. One night I had a nightmare that a glass had gone through my shoulder; I woke up with a pain in my shoulder. So I rang Andrew to see if he was all right, and he said, "Yeah, did you have the dream too?" So we rang John, and he wasn't there. Then his girlfriend came home and rang us and said John's in the hospital after he got a great big piece of glass from a window through his shoulder. That sort of communication comes from being in the same sack together, the same crib together, from being in the same situation together.

We're 42 now and pretty soon one of us is going down—that's the logical order of things. We discuss it now and say, "I wonder which one of us will die first?" and try to make a joke of it, but it's not so much the first one who dies, it's the last the one left alive, who has to write the last sentence and close the door.

I miss my brothers. If Andy or John were in prison I'd break them out, there's no question about it. I used to say nothing could come between me and my brothers, but then my children came along and that kind of changed the equation. My wife and I went through hell with this. When Andrew and John came to visit, she would say, "You think more of your brothers than you do of me," and I had to work all that out. Its like: Well, we're big boys now. It's time we started letting go of each other emotionally.

> *"I was stripped naked, blindfolded, and tied to a metal bed."*

HOW IT FEELS TO BE *KIDNAPPED AND TORTURED*

Maria Pilar, *Age withheld*

IN 1973 THERE was a military coup in my country, Chile. It was like a nightmare. One day you have democracy and the next day you wake up in a different country. We lived under absolute dictatorship. They controlled all media, so nobody knew what was going on in the country. There are no official records from 1973 to 1990.

I was still in high school and just seventeen when they first attempted to have me arrested for being a student representative. I heard that they were looking for me, so I went into hiding and eventually they stopped looking.

I was the leader of the student movement. I felt very

close to poor people and as a middle-class person I felt very privileged. My heart felt connections with the poorer people because they felt true and real. I remember being fifteen years old and seeing very clearly that my path was going to be for peace and justice and, ever since, I have walked that path and been in a position of leadership one way or another.

There was a master plan before the coup to go through the country like a tank and flatten everything that could possible be a threat—even seventeen-year-old high school students—just to signify the government's power. They would do anything that could be seen as a lesson. There was such brutality and viciousness towards the people.

Chilean people are very peaceful people; we were the most economically successful country in South America and were known as the British of South America because we very closely followed the British model, when it was good. We had free medical care, free immunization, that sort of thing, and the coup was designed to destabilize all that—none of that exists now.

The second time I was arrested was the following year, 1974. I was completing my first year at university, studying engineering. I was contacted by old friends from the Left, which was illegal—everything was illegal by then. I heard that there was a strong resistance movement going on underground and people were doing their best to stop the human rights violations, to help political prisoners, to stop the tortures and disappearances.

I was writing articles saying that the spirit was alive and we would return one day to democracy. I was arrested for that. They kept me in solitary confinement and deprivation for three days. The thing that shocked me was that a twelve-year-old girl was arrested at the same time.

They torture people and get them to give up names and find out what is happening. I was terrified psychologically; they threatened to bring in my family, my brothers, my mother. I was kept in a cell with no windows. For interrogation you were taken out blindfolded with your hands tied. You were bashed up a little bit and terrorized. For three days they put lamps on you like a spotlight. Your family doesn't know where you are . . . you just disappear.

After they finished with me, they dropped me in a street late at night without warm clothes or money to make my way home. I think my spirit got a bit broken, and if not, my heart was. I gave up the resistance after that. I was terrified. They did a really good job of terrifying me. I was so demoralized by the experience, particularly seeing that twelve-year-old girl—that really got to me. How can you deal with that? I don't know what happened to her; for all I know she could be dead.

I decided to leave the country for a while, so I went

TO BE KIDNAPPED AND TORTURED

across the Andes to Argentina. I met my first husband in Argentina. He was a freedom fighter there. We went back to Chile together and got married in 1975 and tried to make a living. I went to work as a teacher.

In 1976 our friends were so desperate about what was going on in the country that we just couldn't sit passively, so we decided to join the resistance with two of my best friends. We'd try to campaign for the disappeared ones, trying to release political prisoners and get the government to have some respect for human rights. Almost every household was touched; some member of the family was tortured or arrested or disappeared or killed.

I saw my neighbor's house being ransacked and the whole family being arrested—that was just an everyday thing. They operated with the same techniques as the Nazis. The years between 1973 and 1978 were the most vicious years of the dictatorship, when they killed and tortured indiscriminately.

Eventually the police got to me again and took me to their underground quarters where they had torture chambers. I was stripped naked, blindfolded, and tied to a metal bed. About eight men surrounded me. I was twenty-one years old, so they were being offensive and insulting and having a good time with this terrified young girl. They applied electricity to my genitals and breasts; I don't know how they did it, I couldn't see, but I know it's a torture implement they used.

HOW IT FEELS...

The pain was horrible. It's a form of rape to have those instruments introduced into your vagina and on your breasts, to be naked, tied up, blindfolded, and gagged while all these men are around you having a jolly good time. It lasted until they felt satisfied with the results—you lose track of time, but it was more than an hour. They were all saying, "We'll bring your momma and do the same to her if you don't talk"—anything to make you break down. For me it was instinct for survival that saved me. It's well known that you leave your body in those sorts of moments. Your body is there and your mind is somewhere else, detached.

In another session I had to watch my friend being tortured the same way. Babies were also tortured. I know this for a fact because my best friend had an eight-month-old baby and they and the father were arrested and he was tortured. If he didn't break down, then they started on the wife or the baby. Babies were killed in their mothers' stomachs. If the military had any suspicions of anybody doing anything to threaten the powers that be, they'd take the children to get to the parents. Anything to break down people and terrify them. It was a reign of terror, and they did it very well.

Later the Secret Police came for us. They put sticky tape over our eyes, sacks over our heads, tied our legs and arms behind us, and threw us in the back of a truck. I thought they were going to shoot us or throw us in the

TO BE KIDNAPPED AND TORTURED

water. I thought that was it. After hours of a bumpy ride, we arrived at a place I later found out was a concentration camp, and I was admitted by some military men who asked what the charge was—I was charged with being a subversive delinquent, that was my crime.

I was terrified about what the Secret Police might do. I was put into solitary confinement before I was taken to be interrogated. They give you a medical check-up first to see how far they can take you in torture, how much you can endure. I was in solitary for a month, waiting for my turn to be interrogated. I could hear the tortures going on. There was screaming day and night. I never saw anybody, I could hear them, though. Sometimes it was a child asking for mother or begging for forgiveness.

When I arrived there was something bubbling—they had wells with acid, another way they used to disappear people. They also put people in cages and had people hanging from their feet. For me that was more about psychological pressure, worrying about what the "Gestapo of Chile" would do to me. There were vicious police dogs everywhere, used to attack people, and now every time I see a dog I have to cross the street.

It was a month of waiting and wondering, "Why are they keeping me here for so long, and when my torture comes, will I give up?" But I made a decision that I was prepared to die rather than bring someone else in here as well. I was prepared for suicide. I'd rather die by my own hand, die with dignity. I had two options: the mattress had metal things I could use to cut my wrists, or if that didn't work, I could break a glass. I had plans in place.

I didn't want to die. I had a strong sense of survival. I was in that nightmare and I wanted to get out of there. That was pure hell, knowing that all these human beings were doing these things as part of their normal job and some were enjoying it. At the end of the day, they go back to their wives and kids. How can they finish a shift of torturing someone and then go home to their children?

I could hear the sounds of the city from where I was, with people going to work and to school, and I was locked in there enduring that nightmare and nobody knew. It's how I would describe hell.

Then one day they came and prepared me for a trip. I thought this was it, I was going to be dropped in the sea. By the sounds, I could hear we were going into the city. They took me out of the car and walked me into a courtroom. That was a big surprise. I had a sense of hope that I was going be involved in some sort of legal process.

I found out then that I had been in a famous concentration camp called Cuatro Alamos, which means four cedars. In the court they had a huge list of charges: illegal communication, trying to destabilize the elected government, lots of things. It's amazing how many charges they came up with for a peaceful twenty-one-year-old writing a pamphlet about democracy.

My family was there, but not my husband. We'd separated before I was imprisoned. My mother told me she had been looking everywhere for me. They had denied that I'd been arrested. They told her I'd run away with a boyfriend.

They sentenced me to ten years imprisonment for writing about human rights. I was sent to a women's prison where I had to share beds and tables with women who had chopped their husbands into little pieces, burnt down their houses with the kids inside, and so on. I knew I had been disappeared.

I was there for a month or two and then they transferred me to a prison in my hometown, where I stayed for nearly a year. By this time it was 1978, and Chile was under strong international pressure and had been threatened with an economic blockade. The people of Chile were attempting to create a revolution in peace and democracy.

With money sent from overseas, we were able to get lawyers to defend us, and they found a loophole in the law where you could exchange a prison sentence for exile. So I went for it and it was approved. The day before I was going to be kicked out of the country, they took me to say good-bye to my family. I was ready and packed to go. England had offered me refuge and a scholarship to continue my university studies, but then one of the prison guards came in with a newspaper and told me an amnesty had been decreed and we were all free because of the international pressure.

I stayed in the country for about another five months and joined the resistance once more. I was still being called into the police barracks for interrogation. Sometimes they followed me. You're always looking over your shoulder and your nerves become very frail, and I knew with the file I had that the next time I got arrested I would be shot on the spot.

One day they stopped a bus I was traveling on and it was searched. I thought they were looking for me. I nearly died. I was sweating and shaking, but they didn't find who they were looking for. I knew I couldn't do it any more. I went into a breakdown, paranoia basically, with justified reasons.

I went to London at the age of twenty-two. I had never been on a plane before. I couldn't speak the language and knew nobody. And that was the start of another journey, the journey of life in exile. I came to Australia in 1986. I couldn't wait to get away from England with its cold weather and cold people. Now I go back to Chile every five years to see my family. There are still people in prison there from that dictatorship; some have been there for over twenty-five years. There are still mothers who search the country for their children.

> *"Giving up my children was the hardest thing."*

HOW IT FEELS TO BE DRUG-ADDICTED AND HOMELESS

JANICE SCHILDS, *41*

I BECAME HOMELESS when I was about thirty-seven. I had a family with three children. I had a home, I had clothes on my back, I had money, I had my own business, but I was in full-blown addiction, taking amphetamines. My situation was pretty bad; people were starting to talk and my business was going downhill. My relationship had been abusive and dysfunctional, so I had separated.

I'd started drinking and doing the teenage partying thing at about nineteen. My first relationship was with a dealer. There have always been drugs around me, and it just went on from there. At the start it was just a weekend plaything. By my thirties we were dabbling in and selling amphetamines. I knew I liked speed the moment I tried it. It made me feel normal; I felt good. When there was no money in dope, we sold speed. For nine years I sold the stuff, while for me the drugs momentum was picking up.

I was using during the week but I wasn't IV-ing then. I was either putting it in a drink or snorting it. It wasn't until I left that relationship that the trouble really started. I thought that I could do the single mother thing and run a business on my own. By 1999, I was IV-ing and in full-blown addiction; I didn't realize how bad it was until people started to move away from me. That year I went into rehab for the first time.

Going into rehab meant dropping absolutely everything in my life, and I couldn't do that, so I left and went back home to shut my business down. That's when I let my two children go; I let them live with their father. I told him that within a year I would be fine. Telling him I was sick was very difficult. I kept trying to detox but I would never stay in a clinic. I would always try to do it at home and that would last from nine to thirteen days. I was going crazy, was doing weird things, crying, having fits of anger, isolating myself; my emotions were just out of whack.

My life became totally unmanageable in every respect. I lost my kids, I lost my family, my business, my home. The only thing that I held onto was my car, and

TO BE DRUG-ADDICTED AND HOMELESS

I've still got that today. It doesn't go, but I've still got it. I was never able to get those kids back. In a year I was worse and I was in denial.

I had so much shame that I couldn't face my two kids. I wanted them to see me clean; I didn't want them to see what I had become. Giving up my children was the hardest thing, because I had to admit that I had a problem. I fought it to the end, I didn't think I had a problem, but then, when I really had to stop using, I couldn't.

When I think of the lies that I told; the bullshit that I put my children through—late getting to school, not being there when they came home, being at my dealer's house and coming home an hour late to find them really pissed with me because there was nothing to eat. And there was no love and there was no food on the table and I would just shrug it off and say, "Oh sorry, it won't happen again," but I kept on saying the same thing.

When the kids left I went back into rehab again and I got kicked out. They said I could come back in two weeks if I was still clean. By that time my house was full of misfits. They took over my house. They were using very heavily and I knew I couldn't go back home because when I did I knew I'd use. I was living in my car. I would sleep in car parks near my program. I would park near the Narcotics Anonymous rooms so I could get to a meeting. Then I was staying in my car in front of people's houses—even strangers—because then I'd feel safe.

HOW IT FEELS . . .

I went to the nuns and asked if I could stay. I said, "I need help, I've got nowhere to go." They let me stay there and tried to keep me clean until it was time to go back to rehab. The next time I stayed for seven weeks and then I got kicked out again, because I snuck out and used. I was very willful, very naughty. So then I really had nowhere to go. I was basically in my car, doing the soup kitchens and other charities.

When I used I didn't care where I was or how I was living. When I was trying to get clean I was humiliated to have nowhere to go; it was the depths of despair really, but I was so adamant that I could do it. I really had the desire to stop using. I would go to any lengths just to get clean.

I lived in my car for about six months. It was freezing, I had a duvet and some clothes in my trunk. The people I dealt with were addicts, and they used me and my car to take drugs. So my car took a bit of a bashing at times, but I was so desperate for company and friends I didn't really care who they were so long as I didn't have to be alone. It was sad. I knew there had to be something better.

I did what I had to do to eat, to survive. I had to make money. I had to get some money. I knew there were ways and means. Girls worked the streets. I hated it. The couple of jobs that I did I gave the money to the nuns to look after and said, "Only give me enough for a day, not enough to be able to buy drugs."

TO BE DRUG-ADDICTED AND HOMELESS

When my car became unregistered, I put it in a friend's garage, and then I was completely homeless. I didn't have a lot of friends because I had walked away from most of my past. I was homeless until the Salvation Army called me up and said I could join their Towards Independence program, and that was the one that worked for me. They got me back on my feet, they showed me how to budget my money, they put a roof over my head. First in a block of flats where you stay for seven weeks, then in independent housing for three months, then I was on my own. My caseworker from there still works with me; they've never let go of me completely. She helped me get Red Shield housing and that's what I've got now. I still try to go to NA meetings every day.

To stop using I had to think, "Just don't pick up." It's not easy to stop, but I had nothing more to lose, it was all gone, yet I had everything to gain if I stayed clean. The lowest point was not having my kids, realizing what I'd become, and knowing that I was completely fucked. Many times on the streets I just sat back and thought, "How the hell did this happen?" It happened so fast, it made my head spin. It can happen to anyone.

By the time I got my shit together, which was about twelve months later, and I went home, all my furniture was sold, my clothes were gone, the rent hadn't been paid; squatters had graffitied the walls. All my bank accounts had been drained because, through addiction, I trusted people to go to the bank for me.

It was near-on impossible to get off drugs, off the streets, and back to a normal sort of life. I did a lot of praying. I have a higher power in my life that works for me; you've got to have faith and believe in yourself and that it's going to get better—and it has. I've been clean from my drug of choice for over three years; it got dangerous for me because I was starting to use heroin, and you can pick up a habit within a week. I was popping pills, taking anything to kill the pain. Everything seems OK when you're stoned, but then you've gotta come down off the shit and that's hard.

Life's still not the greatest. I have two more kids from a new relationship I thought was going to last forever. My youngest child was only three months when the bastard walked out the door to a Fleetwood Mac concert and never came back. I've been struggling with that for ages.

I suffer with depression, I suffer terribly, and I'm on medication to give me some sort of balance. A lot of drug takers end up being depressed because they don't know how to cope with life without the use of some sort of mind-altering substance. I don't think society wants to know about homeless people. I see junkies and homeless people on the street and my heart goes out to them. I wish that I had the strength and the facilities to help these people. Society doesn't care; people only want to know what's going on in their backyard.

I get to see my other two kids nearly every weekend now. I've built up a wonderful relationship with these kids. My daughter remembers everything—the lies, deceit, dysfunction, the craziness—but because I've worked to get to this point, they respect me. I never gave up on myself totally. There was something inside me that said, "You're a good person, you can do better, you can do better. You've just got to work, you've just got to push yourself to do it"—and I did.

> *"We were doing everything we could to survive."*

HOW IT FEELS TO
LOSE EVERYTHING
IN A FIRE

Simon Poynton, *34*

THE WHOLE DAY felt a bit ominous—the birds were acting strange and everyone felt a bit odd. The whole energy of the kids and the teachers at school was rather weird. The day was stinking hot and very, very windy. We heard over the radio that there were bad fires around, and we could see the sky was getting darker. It looked like a dust storm, but with smoke everywhere.

I felt pretty nervous knowing the fires were around, because I'd already been through thirteen evacuations in ten years. We'd done it plenty of times, but luckily those fires were brought under control. People who live in the bush know that bushfires are a risk and that you have to be prepared.

TO LOSE EVERYTHING IN A FIRE

We got home from school and pretty soon the place started to fill up with smoke. You could hear a gentle roar in the background, almost like humming. We realized that this one was going to be a big one. No one came around, no police or fire department to tell us to evacuate—we just thought we'd better.

We grabbed a few things. I grabbed my doll Tuesday, my coin collection, and we ran out of the house. I forgot my camera and I cried over that for ages because I'd just gotten into photography and had been given a little Instamatic which was my pride and joy. Losing my camera was a big issue for me and is something I still think about because now I am a photographer.

My dad was a wood craftsman, so he grabbed a few tools while Mom; my sister, Ruth; and I jumped in the car and headed down to the beach. There was a high cliff with a 220-yard walk to the water. We had our two cats, Dudley and Toshka, with us and realized that we didn't have any drinking water for them or for us, so Ruth ran back to the house and got a container of water. My dad was still there and he told her to get out because part of the house was on fire.

We all stood huddled together in a circle in the sea. We spent four-and-a-half hours like that. The water was up to my chest. I held one cat, Ruth held the other, and our parents held us. We had wet towels over our heads because the flames came right down to the sand. The waves kept crashing over us because they were so high from the wind. There were embers flying around everywhere and it was very hard to breathe. I had black boogers in my nose for months after that.

You couldn't see a thing. Every now and then you'd poke your head through the towels, but all you could see was flames glowing. The sound was just ferocious. It was like a jet engine, but constant, just enormous. In the background we could hear explosions as people's gas cylinders from their barbecues and so on were blowing up.

It felt like the end of the world; we were very distraught emotionally. I think I was squeezing half the life out of the cat, and I remember being clawed by it because cats don't like water, but mostly they were very good. They both let out a painful "meow" at one stage and I remember thinking that was when our house was engulfed.

We didn't know whether anyone else was in the water because we couldn't hear them or see them, but later we discovered that another guy was in the water with his dog, about three hundred feet away. It was boiling hot in the water because we were only about forty feet from the fire. If we hadn't had the towels over our heads, we would have gotten our hair singed. It was horrific—something I don't ever want to experience again.

The wind was so strong it was whipping the towels off us and we had to really hold them on. We didn't speak.

TO LOSE EVERYTHING IN A FIRE

It was too loud with the roar of the fire and the noise of the wind, and I think we were all in too much shock. We were doing everything we could to survive. It was a life-or-death situation. But it gave us a sense of security knowing we were surrounded by water.

Four-and-a-half hours later the fire front had passed and you could see it in the distance, heading towards Geelong, though everything around us was still burning.

We ventured back toward the car, but all the stairs in the side of the cliff were burning. Ruth had lost a sandal, so she had to hop while we held on to her. The land was literally red hot after the ferocity of the fire. Afterward, it took a long time for seeds to germinate because it was such a hot fire; they thought it burnt down to about twelve inches.

A fire engine came up the road and told us that one of the local milk bars was open and had sandwiches for any survivors. We went down there to get water and food. It was about 10:30 P.M. Going down the road we came within sight of our house and, when we turned, I said, "Oh look, the neighbors' house didn't survive!" but in actual fact I was looking at the space where our house had been. I looked over and saw our chimney, and that was all that was left of the house.

I'd spent my whole life there, yet there were no features left that I could identify—no landmarks of any sort were left. All the big trees that we knew and climbed, the power pole beside the house, were gone. I'd lost my bearings completely. It wasn't until we went past the empty block that I realized: "This is it." We went back to the store, but it was closed. We got in anyway and just slept on the floor. It was pure and utter emotional exhaustion; it was like: Let's try to rest and see what we can deal with tomorrow.

In the morning we looked out of the window and it was like pictures I'd seen of Hiroshima, with a few buildings half-standing, and all these black sticks that were trees with no foliage. That was exactly what it was like. The ash made the whole ground gray. The amazing thing was that here and there was a house that had survived. That was just amazing. That's when you start to think, "Why me? Why did that house survive and not ours?"

We had a look at what was left of Airey's Inlet, and there was basically nothing—maybe three houses out of several hundred. We couldn't recognize anything; there was enormous open space because all the bush was gone and you could see right through where the trees used to be.

The air was still smoky. We got into where our house had been, and I remember sitting down at the end of where my bed once was. I used to collect marbles, but all I found was a pile of fused glass. The ground was still so hot that the bottoms of my sneakers were melting and my feet were starting to burn.

TO LOSE EVERYTHING IN A FIRE

Everyone was in their room, squatting down to see if there was anything left. My mom found some jewelry, which was all melted together. My sister is planning to make us each a pendant from that gold, like a phoenix coming out of the ashes. It was a complete shock. A month after Ash Wednesday my parents split up. I think the fire was the straw that broke the camel's back. Their history was wiped out in one sense; there were no shared belongings to hold onto any more.

The Salvation Army was fantastic at that time. They were the only reason we survived. They gave us clothes. One guy continued to visit us for about five years afterward. I didn't get any counseling, and it all hit home about two years later when I was about fifteen. I used to sit on the end of my bed and cry with a razor blade held to my wrists. The only thing that kept me going was the thought that people would believe I'd given in.

It took a long time to get over it, and I think every twelve or eighteen months it would surface again and I had to deal with a little bit more. I did that for ten years, I thought. Having dealt with it, I see it as just one of those things, but I do really enjoy my material possessions now, because once I lost them all.

> *"I looked down and there was a hole in my chest."*

HOW IT FEELS TO BE SHOT IN THE HEART WITH A NAIL GUN

Wade Humphreys, *39*

IT WAS ABOUT July this year, and I was building a gazebo. I'm a tradesman. I had the nail gun hooked on the inside of the ladder I was on; it sort of hung down below me. I went down to get it to nail something else, and then I had to go up one or two rungs. I had to reach round and bring the nail gun around the ladder. As I was doing all that, I sort of stepped down, so the nail gun was above me. For some reason I just let it drop from my hand and it made contact with my chest, and as it did I had my finger on the trigger and accidentally fired it.

I've got quite big hands so there's not a lot of room in the nail gun handle when you grab it. If the nail gun had fallen another inch either way it would have missed me, it would have just passed in front of me and I would just have kept walking.

But it got me. The nail hit one of my ribs and went sideways. It went through the front of my lung and out the back and came to rest on the pericardium. I knew that I'd been shot because I heard the bang. I looked down and there was a hole in my chest. There was no blood or anything; it was a sort of red hole.

I went, "Uh-oh. Well, this is it."

I thought, considering where it was, in my heart, and because it was a three-inch nail from a construction gun—that's a big nail—that would be it. I immediately thought about my kids and how I hadn't seen them that morning. I didn't say good-bye to my wife because I'd left so early. I didn't get say good-bye to any of them and I was thinking I wasn't going to see them again. Then I went, "Nah, I'm going to make it through this."

I called my friend Alec, who was working with me, and we got my shirt off and had a look for an exit wound. We couldn't find one, and that made it even worse, I thought for sure I'd end up dead. Any other day I probably would have been working alone, so I was really lucky that Alec was with me.

I didn't feel any pain for about ten or fifteen seconds, then it kicked in as my lung deflated. I have a hard time remembering it now. But the pain started and it hurt, and then it hurt more. It was the worst pain I've ever felt. I've done things like falling out of a car at thirty miles per hour, and had a few accidents with saws and so on, but nothing like this. The pain was intense. When I was still alive after a couple of minutes, I thought there was a good chance that I was probably going to live. I tried to keep calm, but I couldn't breathe, so that freaked me out.

I just stood there and rested against some trestles. I couldn't walk. It hurt more and more and more. Alec called an ambulance, which took about 25 minutes to arrive. It was a long time to wait. We heard one ambulance after about 15 minutes, but it was for someone else. The ambulance guys were great when they got there, looked for the nail, but they couldn't find where it had ended up, either, so they just got me into the hospital.

The first thing the doctors did was to put a hose into my ribs to re-inflate my lungs again. The next day I had surgery to remove the nail. They stuck a camera in between my ribs to have a look for where it was, then went in and pulled it out. The nail was sitting one millimeter from my heart.

I spent five days in the hospital and then went back to work after a week-and-a-half, just doing shorter hours to begin with. I was a bit fried for the first two weeks though. The next time I had to use a nail gun was pretty interesting. I made a concentrated effort to keep my finger off the trigger when I was using it. I was a bit nervous with it.

TO BE SHOT IN THE HEART WITH A NAIL GUN

I think I was really lucky to get through it. I bought myself a lotto ticket, but I didn't win. I figured maybe I'd used up all my luck already. An accident as simple as this, which could kill you, makes you realize how vulnerable you are. It's really changed my view on life and the way I live it. Every now and then I feel myself slipping back and have to pull myself up again. I'm more conscious of the things that I do now, like not getting so worked up about work. I think one of the reasons I had the accident was because I was quite busy and getting caught up in work.

Now I take more time with things and think about them more. I don't stress about work so much. If I don't get a job finished today, I'll finish it tomorrow. I think more about how I speak to my kids and interact with my wife. I might find myself yelling over ridiculous things, small things, but now I can recognize when things are too small to even care about. I really hope that this has changed me for good, because now I take the time to enjoy life more.

> *"There're so many things I haven't done —
> I'm so young."*

HOW IT FEELS TO
HAVE A TRIPLE ORGAN TRANSPLANT

JASON GREY, *Age withheld*

I WAS DIAGNOSED with cystic fibrosis at nine months of age. Until I was twelve I had no troubles whatsoever. Then I started getting stomach pains and stomach problems and I had to go into the hospital to have a bile-duct bypass. Then everything went fine until I was about seventeen or eighteen—that's when I started to notice things were a little bit more difficult. I was getting short of breath, having trouble doing things, and being admitted to hospital more often.

One Christmas, when I was twenty-one, I got really sick and just didn't recover. I was in bed for about three or four weeks. Then I went to the hospital and had some antibiotics for the chest infection, but I had left it too long and it had taken over my whole body. We finally got it sorted out but I never returned to normal health. I just couldn't handle it anymore. So I mostly stayed home. If I went out I had to plan ahead to ensure I'd be OK. I had to know if I would need to walk far, or if I was going to get worn out. I couldn't be spontaneous; life wasn't much fun.

By then I'd had the problem for twenty-five years, and my lungs were deteriorating rapidly. I was frightened about dying. There were times when I felt as if I were on my deathbed and wondered if I would be here tomorrow. I'd think, "There're so many things I haven't done—I'm so young."

It was very frustrating because I'd feel fine one day, and the next day couldn't get out of bed. And I always thought, "Am I going to be OK tomorrow?" There were times when I thought, "Why me, why did it happen to me? I wish I could get rid of it"—all that sort of stuff.

My hospital stays became more frequent and, in 2001, the doctor said, "OK, it's time you thought about a transplant." At that time I said, "No, I don't want a transplant—that's the end of the world for me and I won't consider it." When people said the word "transplant," I thought, "Oh my God, that's the end for me; I'm going to die now." No one ever explained that you have your transplant and you get a whole new life.

They put off talking about a transplant for a while, and then I got really, really sick again and they said,

"You've only got twelve or eighteen months to live; you really need to consider having the transplant." The doctors organized a meeting with my parents and me to talk about my fears and come to some sort of agreement. They gave me a few books to read. Hearing the stories of people who have had transplants does make you think that maybe it is possible for you.

I went to the hospital and I had some tests done to see if I was eligible for a transplant. But they said I had liver problems as well and that the CF was quite advanced, so if they were to try surgery I would probably die on the table. Liver disease is fairly common in people with CF.

They sent me back to my doctor to talk it over, and he said there were some other options; and one was to see a specialist doctor, Keith McNeil, in Queensland, and try for a triple transplant. They hadn't done any over here at that stage, though eight or nine had been done in England, and Professor McNeil had observed a lot of them.

I went to see him and he said I was the perfect candidate for a heart, lung, and liver transplant. He told me to move to Queensland to be near the hospital, so a few months later my mother, brother, and I moved. I went on the transplant list in late August and the waiting began. It was about nine or ten months before I got the call.

That was probably the hardest part. Every day the phone rings and you wonder if it's the hospital. There are days when you think, "Is it ever going to happen? Will it happen in time?" In the meantime I was going back and forth to the hospital. I had a couple of collapsed lungs and some internal bleeding. When you get those episodes, that's when you think, "This is looking bad now."

I was in the hospital with a collapsed lung when the call finally came. I was lying in bed at about eight-thirty one night. Dr. McNeil came in, put his hand on my feet and said, "We may have an organ for you; nothing's confirmed. Hang tight and I'll be back in a minute."

I rang Mom, and a couple of nurses came in and gave me a hug. Mom was there in five or ten minutes. I felt a whole heap of mixed emotions; I was scared and relieved at the same time. I didn't know what to expect or what would happen next. My head was running a million miles an hour.

The doctor came back about half an hour later and said the surgeons were on their way and the organs were organized, although they would take a couple of hours to reach us. I had some blood tests, some tablets, and a scrub-down. By the time all that was done about three hours had passed, and I was ready to go to surgery. It was just like, "It's happening, this is it." I did think that it could be the end.

The surgery took fifteen hours, and when I came out I had internal bleeding and had to go back a couple of times. I woke up in ICU recovery and everyone was

HOW IT FEELS . . .

there—that was excellent. I was sore and out of it, but when you get to see those people standing around the bed, you think, "Yes! I've come through it." It's an incredible feeling.

Three days later they took the breathing tube out and those first few breaths were the weirdest. It's as though you have to learn to breathe again. I'd become so used to taking short shallow breaths that when I got these new lungs I was almost hyperventilating. Having brand new lungs that could breathe was very strange. It took a week or two to get used to it. Being able to take a big breath was excellent. It took me about a month to realize how deeply I could breathe and how good it felt.

I was in the hospital for seven weeks and I was scared at first to come home. I didn't know whether I'd be safe with these new organs. Once I got used to it, it was great. I could hang out with my friends and do what I wanted to do and not have to worry about whether I had enough oxygen. It was life-changing. I still have CF but it doesn't affect these new lungs or liver; it won't re-infect. I used to have physio twice a day. Now I don't have to do so much and that feels fantastic.

Now, after a year, I'm doing great. I haven't had any problems. My brother has CF and so does his girlfriend, and they go into the hospital every now and again, so that's hard. His girlfriend is on the transplant list at the moment.

TO HAVE A TRIPLE ORGAN TRANSPLANT

All three of my organs came from one person. It doesn't feel weird to have someone else's lungs, liver, and heart. The only difference is that I can breathe now. I've heard stories about people who have had transplants and say they take on characteristics of their donor. You know, "I never used to like beer, but now I like it." But I haven't noticed anything like that. My taste buds may have changed a bit, but I think that's just because of the drugs I'm taking.

I don't know anything about the donor; while I feel sad for the family, I'm truly grateful for what they did and what they've given me—a whole new lease on life.

> *"The best thing about having them is that it's five times everything."*

HOW IT FEELS TO
HAVE QUINTUPLETS

Adele Chapman-Burgess, *44*

WE'D BEEN TRYING to have a baby for ten years. We'd done three cycles of IVF (in vitro fertilization) and then tried the GIFT (gamete intra-fallopian transfer) program. Every time a cycle failed, I vowed never to put my body through it again. I thought I was never going to have a baby. Then on the third GIFT cycle, I pretty much knew I was pregnant straightaway.

When they do the egg transfer, they only put in three because that's their ethical quota. I was so sick I couldn't get out of bed. We also knew three eggs had been put back, so my local doctor said I was probably having more than one baby. The thought of having three didn't faze us.

TO HAVE QUINTUPLETS

I had a scan a week later and the assistant said, "How many eggs did you have put back?" We told her and she said, "Well, I can count five. Look, I'll show you." They looked like a little bunch of grapes. Then she went and got the head guy and he came in and was absolutely blown away. There are only four sets of quintuplets in Australia.

I was lying down at the time, which was good. My husband, Bert, was sitting beside me, squeezing my hand and saying, "Five, five! How could this happen?" And he's squeezing my hand harder and harder.

They gave us a photo and we got in the car and drove to the parking lot at Pizza Hut and just sat there contemplating what it all meant. It was way beyond anything we'd ever imagined. We were speechless. I rang my mom and she just laughed and said, "Well, you wanted a baby." I said, "Yes, but I didn't want five babies!" I was in shock, I didn't know what we were going to do, and she said, "We'll just take it one day at a time and deal with what comes; we'll get through this, it'll be all right." I thought, "That's all right for you to say. You don't have to carry five babies."

We have no idea how we ended up with five babies. The hospital had to recall all the nursing staff, doctors, and technicians to go through all the paperwork and check that there were no mistakes. But they couldn't explain what happened. We think either they'd left a couple of eggs in there from a previous cycle or maybe there were already some eggs in the fallopian tubes. It was something they couldn't explain. Those were obviously two very determined babies.

I was worried all the way; at every single stage, at every given moment, there could have been something that went wrong. We worried about conjoined babies, or a child with Down syndrome. I had a bit of a bleed and the doctors were worried that I might lose one. In fact, there is a possibility that we did lose one, that there were originally six.

The doctors said to take it steady with everything and not do too much. They told me if I wanted to have all these babies I just had to be the incubator. And that wasn't a hard decision to make, after waiting ten years for a baby. When such an opportunity comes along, you don't risk anything. We made lots of sacrifices and had a fantastic support system around us.

We live in a little country town in New South Wales, so word got out pretty quickly that we were having quintuplets and the support was fabulous. The town thinks they own the children. Some people were a bit negative, saying, "You'll be lucky if you get one baby out of this." We had to go through a lot of things like that, but we decided only to surround ourselves with positive thoughts.

At twenty-four weeks I flew up on an air-ambulance to the hospital in Brisbane. I had a member of my family with me at all times and people calling in for the whole two months I was there. The staff kept saying, "Keep the babies in; you're doing really well." At about thirty-two weeks they were amazed that I was still going. They had a pediatrician, the head nurse, and two other nurses per baby scheduled to be in the operating room. The whole team was on tenterhooks because they had agreed to be on call for the birth. For two months those guys were living on the edge, waiting for a phone call.

At thirty-two weeks I was getting edgy and also huge. My stretch marks were starting to weep and, until then, I'd had constant movement from the babies, then all of a sudden they stopped. I got a bit worried and stressed, so the hospital decided it was time. I carried the babies for thirty-two weeks and six days, the longest quintuplet gestation in Australia.

I had an epidural, which took an hour to administer. There was a lot of tugging and carrying on and then— there was the first baby, Jack. He peed all over the doctor. All of them came out fine and healthy.

The second one was Louis. For the next I was pleading, "Please be a girl, just give me one girl." And they pulled out Erika, so yes! I got a girl. The next one was India, and the last one was Georgia.

I had to have ten names ready because we didn't know the sex of the babies, but I had plenty of time to do it. Because we are of Aboriginal descent, we chose Aboriginal names for their second names. The babies all got whipped off quite quickly. As soon as Jack came out he was named "Jack Jardie," which means "firstborn boy." The next one that came out, we said, "That's Louis." At first my husband didn't like the name "India," but as she was coming out—this little baby with masses and masses of dark hair—he said, "That's India," so that was it. The other two took a few days to decide.

Not one of the babies needed to go into the special-care nursery or humidicrib. Jack was in an oxygen box for twenty-four hours. He was the biggest at four pounds and half an ounce. India was the smallest. She was just three pounds and half an ounce. My placenta weighed something like twenty pounds.

When I was pregnant, I never thought about holding the babies. You stay away from that. We didn't put a nursery together. Everything was left until we could bring them home. We didn't want to let our barriers down, just in case.

They came home to our little housing-commission house three weeks before their due date. That day I took their armbands off, because Bert and I knew which was which, and by that afternoon we could recognize their cries.

We asked for volunteers to come into the house, and soon we had people in our house 24/7. Our last volunteer would leave at 10 p.m. and the next would be back at 6 a.m. The local social worker fought tooth and nail to get us a nurse for the first sixteen weeks.

There were times when I wanted to walk away from being so full-on all the time. I didn't want to walk away from the babies, but it was very hard never having any privacy. I had forty volunteers a week with us around the clock until the babies were about six months old.

From then until the babies were two, we had volunteers from morning until 6 P.M. Then, people who just wanted to be with the children because they loved them would come on Saturday mornings right up until they were three or four years old. But it got easier as they got older, and the best thing about having them is that it's five times everything. You only have to know how much joy one baby gives to you—then multiply that by five.

When the kids were in grade two, I went to university as a mature-age student and I graduated as a teacher last year. They've just turned twelve and they all get along really well. This year we've got two school captains, one of the girls is on the Student Representative Committee and two others are house captains. It's starting to sink in to them that they're pretty special.

I believe God dishes out to people what he thinks they can handle.

> *"The wind was a roar, it just screamed. I've never heard anything like it."*

HOW IT FEELS TO
BE CAUGHT IN A CYCLONE

PETER EVES, *44*

I'D HAD A small operation and was watching the news in the hospital. I'd seen Cyclone Vance start off as a small low on the weather chart. I watched it for a few days, then eventually I said, "I'm out of here," and discharged myself to go home and pack up our gear. My wife Vicky and I were operating a salvage business at the time. I knew that I wouldn't be able to do too much after the operation, so on the way home I picked up a hitchhiker and offered him some work helping me to prepare.

With cyclones you have good warning; they don't just pop up in the middle of the night. This was the worst one we've had. We've been here in Exmouth, Western Australia, for ten years, and there have been about four that have come close to us. The longer you live here, the better the odds get in their favor.

A lot of people pack up and leave, but what do you do then? At that time we had no prediction of where the cyclone was heading. It was the luck of the draw. It could have gone on the outside of us; it could have turned off and gone elsewhere. It's pretty frightening knowing a cyclone is on its way.

A lot of people were very nervous and very jittery and for quite a time afterwards the wind would blow and you'd hear that whistling noise and the hairs would stand up on the back of your neck. The night before it hit, we were listening to news reports and tracking it on maps. We were getting rain and winds of about forty miles an hour, just strong enough to be uncomfortable.

Exmouth was originally built by the U.S. Navy; they build their houses with a lot of concrete—concrete walls and concrete roof. Fortunately we've got one of those on either side of our house and two behind us. One was vacant at the time; it was a holiday house belonging to the police union. I got in touch with the police and asked if we could move there for shelter if things got bad, because we were in a thirty-year-old fibro-and-timber house. Like the three little pigs, you always go somewhere stronger.

My wife Vicky and the two kids moved over the night before, and I stayed home till the next morning.

HOW IT FEELS . . .

There's not much sleep. You're listening to the forecasts and thinking about the cleaning up you didn't get time to do. I woke up at about 5 a.m. and the wind was really starting to scream, so I grabbed a few more things and started moving them across to the other house. By 8 a.m. she was fairly humming.

The eye of the cyclone passed here about 10 A.M. Then the wind changed, the eye crossed very close to us, and the wind came from a different direction. As the eye came closer we got what's called a whiteout. The rain is driven so hard that you could open the back door but see only about ten feet ahead. It's like a snowstorm, but it's just rain traveling hard and fast.

The wind was a roar, it just screamed. I've never heard anything like it. You're hearing things banging off the wall and off the roof. Our house was next door and we heard a couple of big thumps, and I thought we'd lost the house.

The sky was just totally black; it was almost like night because the cloud is so intense there's no light. When the eye of the cyclone hits, you get a bit of light—it comes up like a dawn glow for a short time. At that stage it created the strongest winds ever recorded in the Southern Hemisphere; the official speed was 166 miles per hour.

At about 2:00 P.M. it subsided enough for us to go outside. We went walking around, looking at everything. The damage was quite incredible. There wasn't a leaf in town—well, not on the trees anyway. They were plastered all over people's walls and floors. Our house, which survived, was full of them because the windows and doors had blown open, so the inside of the house was full of debris. Instead of foliage, the trees all had tin leaves—bits of scrap metal and aluminium.

There were no deaths. There were a few injuries, but no one was seriously hurt. One lady got stuck for six hours underneath her house after it collapsed, and she sat there singing hymns all night. There were quite a few houses lost. We lost our business premises and a lot of gear; we had to start from scratch. The cyclone blew all the walls away; everything that was inside got thrown around. The SES (State Emergency Services) turned up to help us with the cleanup, and did in a few hours what would have taken me days on my own. All that Vicky and I could do was hug each other and cry with relief and sadness.

We had lots of personal items stored at the business—photo albums, books, and so on, which we lost. Vicky lost her silver tea service and was pretty upset about that, but one day she was out walking and found the sugar pot lying in the mud. The other premises, which I was building, lost all its doors. We found one huge door five hundred yards away. One set of doors we never found at all. They're probably out at sea.

The town, which has about twenty-five hundred people, lost its water supply. Over here we've got American-style water hydrants that stick up from the ground. A van got blown along the street and knocked a hydrant over and that drained our two tanks of water. We lost our electricity and phone lines too, and didn't get anything back for about two weeks. We had no services.

Everybody just went camping in their houses. It was still raining, so you filled up whatever you could with water and used it to flush the toilet and to drink. Everybody just showered under the run-off from the roof. It was back-to-basics, but no one really worried.

The biggest drama was getting the kids away. It took a couple of days to get that organized. Because we had no water or power it became a health issue, so we sent them to their grandparents. We had water and supplies flown in on Air Force Hercules, and they then took people out—mainly tourists and women and children—but it took days. The rest of us stayed in town because we had to clean up.

After a few days, the Council went around with a big skip—we called them the "Pong Squad"—and made everyone empty the contents of their freezers because of the danger of rotting food.

This sort of thing really brings out the best and worst in people and a community. The F-word really became the word in town. We were all just walking around, looking at the devastation, saying, "F——!" Even the little old ladies were doing the crow call. We were all stunned, with our eyes like saucers, but it was quite hilarious how all we could do was swear. Some people were so shell-shocked that they packed up and never came back.

> *"It's all about prolonging your life,
> rather than finding a cure."*

HOW IT FEELS TO BATTLE CANCER

Mark, *35*

I'D LOST LOTS of weight—about twenty-one pounds over eighteen months—but it didn't bother me too much. I thought it was because I was doing more work in the garden. For most of 2003, there were times when I was feeling lethargic and slightly flu-ish, but I never investigated why.

I also had shoulder pain. I thought it was maybe from using the computer too much. I told my GP in June, but he didn't make any connection with the symptoms of liver cancer. There weren't many side-effects. It was originally bowel cancer that spread to the liver, so bowel movements had changed over a long period of time, but there was never any direct sign that something was really bad, no blood or anything.

In the middle of October, for about four days in a row, I had a pain low on my left side. I went to the GP and he checked me out more thoroughly. He ordered blood tests and an ultrasound. The blood tests didn't reveal anything, but the ultrasound showed that there was something in my liver, something quite large on the right side. I had a CT scan straight away and that revealed a seven-centimeter tumor in my descending colon which was almost completely blocking it off, and also a four-inch-in-diameter tumor in my liver which occupied the whole right third of it.

I was worried, though I still didn't think it would be cancer. But when I went back to the GP he started using these big words and I knew it was more serious. I had tears in my eyes when he was talking to me, and driving home I was very emotional and teary. Having to break the news to my wife and parents was hard.

It was all very, very quick. Only about two weeks from the tests to my being in the hospital. There wasn't time to be mulling over everything, you're just in the doctors' hands and you do what you're told and go along with everything.

I thought that this could be a death sentence because there was a tumor in the liver as well as the bowel—

though I knew that the bowel cancer was bad and had probably been there for quite a while judging from the size of it. There was also a worry about where else the cancer could have spread that wasn't showing up on the scans.

I was referred to a surgeon to have a colonoscopy, which wasn't successful because they couldn't get past the tumor. I had surgery a week later to have the bowel cancer removed. I was in the hospital for fifteen days because there was a complication after the initial surgery. My intestines had developed adhesions where they had been flattened, like a hose squashed together, so my digestive system wasn't working and I had to be opened up again a week later to fix that up. I lost a lot more weight and dropped down to about 144 pounds, so I looked pretty gaunt and awful. I felt sick and in pain the whole time I was in the hospital.

It was a very painful and slow recovery, and there was all that worry about not knowing the future. I focused on recovering from that surgery and blocking out any thought of the next operation in February. I was also told I'd be having chemotherapy afterward. I virtually couldn't do anything for two months during the recovery phase. It's extremely debilitating being cut down the middle; all those muscles control just about every move that you make, and so I couldn't do much.

The extended family's worry about what was happening didn't get relayed to me until later on. I guess they were trying to be strong and not upset me. You do get upset talking about it, but actually talking would have helped. I would have liked to know how they were feeling, or how my being this way, or maybe dying, was going to affect them. I wanted us all to be honest.

Some family members have been able to talk to me really truthfully, and that's helped, but others have reacted by not knowing what to do. None of that bothered me; how people reacted to me didn't upset me. It was more my own feeling of "How many years have I got left? Or is it less than a year?"—that sort of thing. I wondered how long I had to live and whether there was any point planning anything, or whether I would miss the kids' growing up. It's the things I would lose by dying that upset me more than what other people said. Their being upset didn't upset me. Talking about how they would deal with this loss in the family was fine.

After the liver surgery I was in the hospital for eight days; then I had to recover from that and start chemo. Just before the chemo started I had another CT scan which showed some new spots in my liver and in my lung. That was really, really devastating. That was my lowest point. I went to see the surgeon and he mentioned other treatments, like lasers and so on. But it didn't sound very promising.

Already going through the two operations, and recovering, and then being told there was something else—

that was certainly the worst thing I've been through. It was worse than the initial diagnosis. Going through all the pain and crap with the operations is one thing, but then to know that the pain and stress are going to linger on, and to have that anxiety on an ongoing basis is pretty upsetting.

I started the chemo, and now I've done five cycles. I had another CT scan last week with a new machine that allows them to see 3-D images. This showed that the tumor in my liver has gotten smaller and it might not be what they thought it was. There was also nothing in my lungs, so either the chemo was doing the job or you could even suspect that those images might have been scar tissue from earlier infections. This was more positive news. It gives me hope that the chemo is working or that the other things aren't too serious yet, so that I have more time. I'll continue with the chemo and hopefully I'll have a bit of life left after that.

Even though the bowel cancer may have formed up to three years ago, and the liver thing was probably there for a year-and-a-half, I have not felt like blaming the doctors for failing to pick things up earlier, nor have I felt any anger. But I have had the "why me?" feeling, because I don't smoke and I don't drink; I've led a pretty calm sort of life, and I eat sensibly—so it just seems really unlucky and unfair. When I talk to the oncologist, the way he describes things is that it's all about prolonging your life, rather than finding a cure.

I don't think cancer has changed me a great deal. Sex is something that bothered me. Everything changed; you couldn't do what you wanted to do and felt ugly because of the scars and soreness. So that put me off. My wife and I have talked about the future—like, if I do die, where I want to be buried and things like that, and how she'll manage financially. We want to have as much family time as possible, short holidays and things like that.

I have a lot of sadness about missing out on things in the future, missing out on life with my wife and kids and friends; thinking about things around the house and garden that might take years to do—and what's the point of that? You lose hope and it's hard to see yourself in the future.

"He shot my sister three times in the head."

HOW IT FEELS
WHEN THERE IS A MURDER
IN YOUR FAMILY

DAVID, *Age withheld*

IT HAPPENED IN 1973, two days before my sixteenth birthday. My sister, Maryanne was twenty-four when she was killed. She had been married for three or four years, but there was a growing tension around her husband's sexuality and, as a consequence, they split up, though they remained on good terms.

A friend of a friend came to stay with them shortly before they split up; Maryanne and he became good friends and eventually lovers. He was a heroin addict and he either introduced her to heroin or promoted its use. But he was quite abusive, physically and verbally, and as a consequence they were only together for about three months before she told him to leave.

He continued to pester her, though, and she felt quite vulnerable. But after he hadn't contacted her for a couple of weeks, she believed that it was finally over. Then he got himself high and went to her house with the intention of killing her. He had a gun with him, wrapped in a cloth. He put it in an outside bathroom at Maryanne's house. Because it was a weekday, he assumed she was at school teaching. He must have gone into the house just to have a look around, but discovered that she was asleep in bed with the flu. So he went back out to the bathroom for his gun. He shot my sister three times in the head and then shot himself in the heart.

The landlord came around at 5 p.m. that afternoon. He and Maryanne were good friends, and the back door was open, so he walked in. He found them in the bedroom with her lying in bed looking quite calm and peaceful because she was shot in her sleep, and him spread across the bed. It was obviously a horrific scene.

My family was told right away but, because I was working on a dairy farm out of Melbourne, they didn't really know how to tell me. They decided one of the family should come and get me rather than tell me over the phone.

The morning after Maryanne was shot, I was on the farm milking at six in the morning and a Mercedes pulled into the driveway. This was a most unusual thing on a dairy farm, for such a flashy car to be there. I thought it must have been someone for the boss, so I just kept milking. The next thing, the boss tapped me on the shoulder and said, "David, your sister's here." It was my younger sister. I went over and gave her a hug. It felt rather strange, because the family had never visited me on the farm. She took me into a room next to the dairy and told me that my elder sister had been shot and killed. It was just the most surreal experience to be told in that environment—in the dairy with all the pumping machines going, and the cows—to find out that your sister's been murdered. The fact that she died a violent death was abhorrent.

I couldn't stand it. I just went into shock. I was waiting for my sister to tell me it was a joke. It was 6:00 A.M., the dawn was just breaking, it was freezing cold, and there we were—me with my ten layers of clothes and gumboots, cows mooing in the background, the dairy going full pelt, and my sister telling me that our older sister was murdered yesterday.

I had this feeling of absolute disbelief. I remember trying to pack up my bag to leave and my sister trying to be practical, saying, "David, where are your underpants, where are your socks, what do you want to wear?" And I'm thinking to myself, "I don't care if I wear underpants and nothing else," because making decisions like that seemed to have no meaning at all. It just had no significance.

WHEN THERE IS A MURDER IN YOUR FAMILY

The funeral was about three days later. My mother was absolutely distraught. She wondered whether the way she had dealt with Maryanne's boisterous personality had had an impact on her death—or at least on her decisions. Mom's never been the same since it happened. Even now, thirty-one years later, she is still given to more volatile emotions than most people, both sadness and joy. It's interesting, the scar that the murder has left on her. I don't think you could wish upon a parent any more terrible or traumatic incident than losing a child. My mother just never really recovered. In those days counseling wasn't routine, so she was never offered any support.

The man who killed my sister, his parents were Canadian. The police told us that they had to send a report over to Canada and his body would be sent back there. My mother asked if it was mandatory that his parents know he murdered her daughter and the police said not necessarily. So my mother insisted that, for the parents' sake, they not be told. To this day they have no idea. She said there was nothing to be gained by their knowing; it would only make them feel more distraught. It was such a generous suggestion.

Because Maryanne was nine or ten years older than myself, she was like a second mother to me. She was intriguing because she was from the seventies art world. I used to stay with her and my brother-in-law quite often;

it was like a home away from home, an escape from the tumultuous atmosphere at our house, because Dad was an alcoholic and his emotions were all over the place.

My family hadn't told me that the man who killed Maryanne was a heroin addict. When it did come out, there was no suggestion that Maryanne had also indulged. The last thing I would have believed was that she had been hooked on drugs or would have used a drug that I considered to be incredibly dangerous and counter-productive.

About a year after her death, my younger sister Robin told me that during the autopsy they found track marks on Maryanne's arms. I remember sitting in the back of Robin's car and I cried from the time we left Geelong till we got to Melbourne. I just couldn't believe that my sister had used a drug like that—and that there was physical evidence she had used it more than once, indicating it wasn't just experimentation. I then found out that one reason she broke up with the man was that she wanted to stop using drugs. I'm sure she wanted to change her life, but he put a stop to that.

Her death is still a source of great sadness. Maryanne was a really great artist. I've no doubt that if she'd stayed alive, she could have been famous. She really was incredibly talented and very well-respected as an artist. That alone fills me with sadness, that level of talent being wasted. She was volatile but, at the same time, amazingly joyful. She had a fiery spirit and was a very funny person. She's often in my thoughts, and some of my personality traits are reflections of hers.

She would have brought joy to a lot of people and got a lot out of life herself. I often think, "What would she be like now?" It's hard to imagine what she would have been like at fifty-four. I feel great anger towards the man who killed her. I find it difficult to forgive him for not exercising a greater level of control over his anger.

My mother is glad he killed himself because she wanted to kill him. I would have been after him for sure. Small as I was, I certainly would have been looking for him.

> *"One thing they tell you is that you've got to give up thinking."*

HOW IT FEELS TO BE BRAINWASHED BY A CULT

Luke, *34*

IN 1999, MY wife Sally's doctor advised her to spend a few days with this group for respite care. She was suffering postnatal depression after having our second child.

The doctor didn't know anything about the community. She thought they were just really nice people. She didn't know they were actually a cult—and at that stage neither did we.

After the second day there my wife believed she was a sinner going to hell. The only hope she had was to give up everything and live there. She was ready to leave me and join. She didn't know anything about the group, but it filled some deep need within her. She was looking for something to make herself feel valuable.

I'm not sure if she would have taken the children with her or not at that stage. But the cult told her she had to come home and "win me over"—to get me into it, too. That process would take about six months. Unbeknownst to me, she told them I'd be bringing in about $400,000 to the community, so they were very keen to get me in and really used the big guns to persuade me. She was a bit unhinged—we had nothing like that amount of money.

I thought we had a fairly good marriage. I didn't know she was so desperately unhappy. She didn't appear to be, she didn't tell me that she was, but obviously the group provided something that was missing in her life. So eventually we joined. I gave up my family business, giving it back to my father. I was doing contract manufacturing, producing chemical blends for customers. We sold our house and paid off the mortgage and had a big garage sale to sell all our possessions. Whatever was left we gave away. We went into the community with clothes and a few personal items.

We gave them around $12,000 to $13,000, which was everything we had. I convinced myself that we were doing the right thing. A lot of what they said made sense, but the big motivation for me was that I was pretty sure I'd lose Sally if I didn't join.

My parents were thoroughly pissed off because I didn't really talk to them about it. I sprang it on them, which was a sign that I wasn't really comfortable with the decision.

Once you join you have no car, no money—if you want to do anything, you have to ask. If you want a new pair of underpants, you've got to ask. Each day you'd have no idea what you would be doing; you'd be rostered on for different jobs in the community. Being a qualified teacher, I did some teaching in the mornings, and in the afternoons I would do manual labour—cutting lawns, digging trenches. I did a lot of driving—picking up and delivering things.

My wife and I moved into a room about sixteen feet by eleven feet wide, and then we had an extra room added for the kids. Everything else was shared by the whole community—bathrooms, toilets, and kitchen. Every day the community would have breakfast together; whoever was working together would have lunch as a group; and we'd all have dinner together.

Women took a very subservient role, except for those they called the "Wise Women." They had power. All the women had to dress modestly. They generally worked in the kitchen or the laundry. That's about it. My wife didn't have a problem with that.

If you were single, you weren't allowed to spend much time with individual women. If you were interested in possibly marrying one of them, you'd have to ask permission to go on what was called a "Waiting

Period," where you were allowed to spend an hour or two each day talking together. You weren't allowed to hold hands or have any other physical contact. Once you got engaged, then you were allowed to hold hands; the rules were very well-defined.

I wasn't supposed to have sex with my wife or even sleep in the same bed with her during her menstruation because she was "unclean." There were clear guidelines about what was considered godly and what wasn't. Male masturbation was out, unless you were a single brother, and then you could masturbate if you weren't thinking about any woman in the community. That was like a deadly sin. Female masturbation was totally out. Women were considered to have no need for that whatsoever.

Three months after we joined the community our youngest child was born. They wouldn't allow doctors, but we had a midwife. Four weeks after the baby was born he came down with RSV bronchiolitis, which is life-threatening for a newborn. At this stage we were still in contact with our own doctor, and she said we needed to get him to a hospital immediately. We told the elders, and they said, "If you want to take him to hospital, we support that decision. However, if you have enough faith in God to heal your son, then we'll support that decision as well."

All of a sudden it came down to a test of faith. I said, "I don't have enough faith, so I'll have the car and take him to hospital." We got him into the emergency room just as he stopped breathing. He would have died if we hadn't taken him.

Despite all that, for the first year I was happy. There were some positives, but there were a lot of negatives, too. Getting rid of all of my possessions was great. It was so liberating. I never really liked having much stuff anyway. Living with a bunch of people who are really passionate about something is also a great experience. You'd never get that in any workforce—so many people working for free and really striving toward a common goal. However, the problem was that goal was a lie. It was not a true common goal, and I think people were getting duped.

It was just complete brainwashing, but at least I kept my ability to reason intact. One thing they tell you is that you've got to give up thinking, though they don't say it in such a direct way. I have quite an analytical mind, though, so I just kept a tally of all these things that didn't add up. I kept a tally of what they told me and then what actually happened.

In hindsight, I was flabbergasted they were able to have such a hold over me. If you had asked me, five years ago, who would be the least likely person I knew to join a cult, I would have said me. No way would I do that. But they're very subtle; they're really lovely people, they look in your eyes and they speak with conviction, and you want to believe them. They don't tell you everything; they just tell you the good bits. They tell you what they think you want to hear. One of my big problems with them was that I asked very specific questions relating to their being a cult, and they lied to me.

It was very oppressive for children. The discipline was horrendous. It's all based on the Bible, and they give very good arguments for why it's necessary. I'm quite ashamed that I didn't leave earlier, particularly after one incident out there. We'd only been there for a month or two at that stage. I left this man in charge of my son. He asked my son to come to him and the boy didn't, which is fairly normal three-year-old behavior, so the man smacked him six times on the hand with a balloon stick, which is a thin bamboo rod. It's very painful.

He put my son down and asked him again to come to him, and he didn't. I asked how many times he did this to the boy, and it was about ten times, and each time he hit him. Then he told me, "But finally the child submitted."

I'm sick when I think about it now. But I was in a different place then. At the time I didn't know what to do. This man had been there a lot longer than I had and therefore knew more than I did. Particular people were allowed to beat the children. If you were left in charge of a child, most times you'd be allowed to discipline them physically.

I was never aware of any sexual abuse of the children; that was one thing I couldn't have accepted. However, in overseas communities of the same cult, quite a few children are bringing legal cases against the elders and others for sexually abusing them.

After the first year, I started asking questions. I wanted people to tell me particular things and why they believed certain things.

What I found out was that a lot of people there didn't believe in the teaching; they just went along with it. But I didn't join this group just to go along with it. I joined it because I believed in what they told me— that it was good to live with one another, to love God by loving each other and by laying down your life to do that. It's like *Animal Farm*: everyone's equal, it's just that some are more equal than others. But it's not even close to being equal.

In November 2000, Sally and I were sent to a community in Canada to work. That was when my eyes were really opened, because this was a much older, larger community, and it was quite dysfunctional. It was there that I could see the group for what it really was. After about six months there, I lost all interest in the group and just started breaking myself out of it. Sally wouldn't say a word when I expressed my concerns to her, and as soon as I finished talking to her she'd go off and tattle on me.

TO BE BRAINWASHED BY A CULT

We were over there for 9/11, and one of the leaders got up one day and said we had to be more like the people who flew the airplanes. Not that we should go out and kill anyone, but we should be so focused that nothing else mattered. I said to Sally, "That is what scares me. We cannot just be blind robots. We've been given brains and we need to use them." I suggested we should go home and revisit our decision based on what we now knew about the group. She went and told the elders and they kicked me out for threatening her, because I had said if she didn't come I was going to leave and take the kids with me.

It was pretty stressful. I only had a hundred dollars given to me, and they dropped me and my son off at a hotel which had pornography on the front counter and signs advising guests to double-lock their doors. I got in touch with my family; they wired me some money, and my brother came over with a plane ticket. I still didn't realize at this stage that they were a cult. I just didn't believe in their philosophies any more.

When Sally and the other kids got back, I arranged a meeting with her. She turned up with all the kids' clothes packed—everything, even though it was just supposed to be a weekend visit. She was still breastfeeding our youngest child at the time. When I got the baby seat out of the car, I saw that she had a rod there to discipline him on the drive. This filled me with rage and, at the same time, shame for what I used to do as well.

We went for a walk while my mom looked after all the children. I tried to talk to Sally and told her why I left; I talked about our history, and she never said a word. She wanted to be with the group more than with me or her children. She left without saying good-bye to the kids.

I then tried to arrange visits between her and the children, but she wasn't interested. We had a few phone calls in 2002, and that's been it. I have no idea where she is now. It would have been easier if she had died in a car accident, easier to explain to the children. It's very difficult to say that she has a mental illness and that people are lying to her. The kids are now aged eight, six, and four.

Every mother must know it's not OK to just abandon her children; it's the worst thing a mother can do, particularly for my eldest daughter. She was in tears one night and said, "Mommy must hate me." I've been very diligent in keeping an eye on them and getting them counseling when they need it. I've now been remarried for six months and that's been really good for the kids.

I got thrown in the deep end caring for three kids, but I love it and I always had the support of my family. I just focused on the kids and enjoyed being a father. It was a relief to have my life back.

"Being on fire is a weird sensation."

HOW IT FEELS
TO SET YOURSELF ON FIRE

BRIAN CONCANNON, AKA MR. INFERNO,
Age withheld

I'M AN EXTREME daredevil stunt performer, but because I'm not accredited with the Media, Arts and Entertainment union I can't actually do stunts on film sets. A lot of accomplished stuntmen frown at me because of the way I have gone about making it as a stunt performer in the face of adversity. I've tried to get my "colors," but I've sort of hit brick walls. When I was doing my stunt course years ago I went in a bit too reckless and ready to do anything. I would watch experienced stuntmen do a stunt and I would give it a go myself straight afterwards without the necessary practice or experience. Unsurprisingly, I got shown the back door and after many unsuccessful attempts to get into other stunt courses, I decided to pursue my dreams of becoming an accomplished stuntman by doing my own training. This was the greatest skill I had in life and I wanted to make something of it.

This attitude made the other stuntmen fearful of me; the way I paid no attention to safety or proper procedure. I can see why they thought I was an accident waiting to happen and shunned me. The same thing happened to me at skydiving, I freaked out the "sky gods" community at my local drop zone with my reckless attitude. After a couple of hundred jumps they barred me because they thought it was a matter of time before I bounced (died). I wish I'd acted differently—more responsible and competently. I really miss skydiving, it's a great sport and now I realise that you can't be complacent or too gun ho, it just doesn't wash.

At first thrill seeking and daredevil stunts were a hobby, but in the last couple of years I've set my sights on turning it into something that can bring me financial benefit. I've got these talents and I want to be acknowledged for them. My skill is the ability to do different stunts while on fire. For most stuntmen just being on fire alone is their stunt, but I've got to the stage where just through trial and error doing backyard stunts, I've gotten to a level of confidence where I'm starting to add other stunts with my fire burns, like walking a beam or skateboarding etc while on fire.

TO SET YOURSELF ON FIRE

Being on fire is a weird sensation. The sound alone is quite scary and the heat and flames burning around your face is terrifying. The knowledge that you're red hot on the outside and that this is a totally wrong thing to be doing is a freak out, much like jumping out of a plane or base jumping.

Depending on how long a burn I'm doing I'll have more and more, multiple layers of safety clothes on which makes you very claustrophobic and feel like your cooking inside. Usually I just use gas as an accelerant. If I'm jumping into water I sometimes use a flammable glue because it leaves no pollution.

I've done fifty-six fire burns. My longest was two minutes, 18 seconds. The world record is two minutes, six seconds. I might have officially got the new world record, but because I didn't have two professionals to witness and verify it, the *Guinness Book of Records* couldn't count it. I've got the video but that didn't count. It's considered one of the most extreme stunts in the stunt industry along with the highest high fall record—so I know I'm onto something.

I've copped a hard time being in the media spotlight, people say I'm setting a bad example and just taking stupid risks which I would have to agree with, but the way I see it I'm performing my stunts for my own experience and future financial benefit because I am one of the best in the world at what I do.

HOW IT FEELS . . .

It's no use applying for permits or authorisation, though if I was in a position to successfully apply I would. I try to perform in areas where the risk of being noticed is minimal most of the time. Occasionally I will do a stunt that gets noticed and emergency services respond which takes them away from where they are really needed. I sincerely regret this and know what a pain I must be to the authorities.

I know that what I'm doing is risky; cutting corners without the proper safety requirements, but I have never experienced a really dangerous situation and developed a false sense of confidence. Well, not until I attempted to ride a bike off a toilet block roof while I was on fire and land on a landing pad of mattresses.

Suddenly the police turned up and I stupidly rushed ahead without being fully prepared or any practice runs. As a result I set myself on fire and took off in the wrong gear, didn't get the right speed up and missed my landing pad, hitting the ground hard and breaking my wrist and nose on impact. I had trouble putting myself out. I couldn't wrap a wet blanket around myself because of my shattered wrist. Luckily a brave policeman came to my aid and repeatedly told me to lie down and poured a bucket of water over me, saving me from severe injury or death. When I have more money I want to get that policeman a case of the finest scotch for acting fearlessly in saving my life.

That stunt was a real wake up call for me. I had plates put into both bones in my wrist, a broken nose, a tooth knocked out and second-degree burns to my right arm and shoulder. I spent three days taking up a hospital bed for a senseless, stupid action. I was lucky to escape without horrific burns or losing my life.

I have performed a few stunts since the ill-fated bike jump and have regained my confidence. I've learnt never to rush a stunt again. Danger is a rush, but rushing danger is a death wish.

NOTES

SINCE THIS BOOK was originally published in Australia, all references to dollars ($) refer to Australian dollars. Metric measurements have been converted to English units where appropriate.

Have you got an amazing or compelling story you'd like to tell? We'd love to hear it, and who knows? It may end up in a book some day.

Please e-mail your story to Michelle Hamer at howitfeels_641@hotmail.com

If you want to know more about our interviewees or their experiences, here are some Web sites and book titles that may help:

How it feels to be abducted by aliens
   www.users.tpg.com.au/ufoesa/

NOTES

How it feels to survive bacterial meningoccocal disease
Amanda Young Foundation
www.amandayoungfoundation.org.au

How it feels to be lost at sea
Derryl and Aileen Huff are now raising money for the coastguard and Westpac Rescue. To make a donation contact: *nelltarly@felglow.com.au*

How it feels to die and come back
Read more about Ken Mullens at
http://home.iprimus.com.au/gotztass/km/nde1.htm
Contact him at graken234@bigpond.com

His self-published books are *Returned from the Other Side* and *Visions from the Other Side*.

ACKNOWLEDGMENTS

IT HAS BEEN a privilege to step into the lives and hearts of so many courageous and positive people. This is your book, your stories, and I thank each of you for being so honest and open in sharing your story.

Thanks to Alice Ghent and Veronica Ridge at *The Age* newspaper, which ran the first *How It Feels* series, which inspired this book.

Thank you to the beautiful Tracey Caulfield, who drove me to my first appointment with the publisher, helped me come up with great ideas, listened to me moan on bad days and babble on good days, cooked food for my kids when I ran out of steam, laughed at me when I decided to quit (every time!), and continued to be the best friend that she has been to me since we met at fifteen.

Harley, Ruby, Darcy, and Oliver—I love you all. Dream big and don't let go.

ACKNOWLEDGMENTS

Kell and Ads, thanks for your ideas and support. Love you both. To the future!

Thank you to Cathy Smith at Lothian Books for always being wonderfully positive and upbeat.

And lastly, thanks to co-publisher Geoff Slattery and Selwa Anthony, my agent.